# ASTRONOMY

## Selected Topics

by

Dr. Charles H. Grace

Order this book online at www.trafford.com
or email orders@trafford.com

Most Trafford titles are also available at major online book retailers.



Printed in the United States of America.

ISBN: 978-1-4269-3442-1 (sc)
ISBN: 978-1-4269-3443-8 (hc)
ISBN: 978-1-4269-3444-5 (e)

Library of Congress Control Number: 2010909742

*Our mission is to efficiently provide the world's finest, most comprehensive book publishing service, enabling every author to experience success. To find out how to publish your book, your way, and have it available worldwide, visit us online at www.trafford.com*

*Trafford rev. 7/21/2010*

**North America & international**
toll-free: 1 888 232 4444 (USA & Canada)
phone: 250 383 6864 • fax: 812 355 4082

# Table of Contents

# Chapter 1
# Earth

## The Amazing Global Positioning System (GPS)

Many people in many fields use the Global Positioning System

1. Military officers. The U.S. Dept of Defense, which developed the GPS, can keep track of troops, target cruise missiles and precision-guided bombs, and detect nuclear explosions anywhere in the world.
2. Navigators of cars, planes and ships rely on GPS. Glider pilots and mountain climbers also use it.
3. Surveyors use GPS to locate boundaries and surveying markers. Specialized road construction devices are even capable of manipulating blades and buckets of equipment for grade control by GPS.
4. Some scientists and engineers need precise time references. The GPS time signals are used to set time-code generators and NTP (Network Time Protocol) clocks for synchronizing widespread computer networks.
5. Geophysicists and geologists. Example: GPS reference time signals can record the arrival time of earthquakes at various recorder locations very accurately.

6. Mobile phone engineers. Mobile phone locator signals that use conventional radio are not nearly as accurate as GPS locators are, so mobile phone designers can get by with less powerful transmitters by using ancillary GPS.
7. Employers of traveling employees can keep abreast and help them. For example, a drug company sales manager can see that one of his salesmen is presently calling on a CVS drugstore in Lakewood, Ohio.
8. Husbands and wives can track spouses and others.
9. Care-givers can monitor elderly people and know where to look for them by contacting a GPS device carried in their clothing. Pets that are equipped can also be located.
10. Parents keep track of children. It works so long as the kids don't leave their GPS device somewhere else.
11. Car and truck fleet owners can track their vehicles. GPS can also control harvester vehicles as they work in fields.
12. Police can find stolen autos and fugitives. Police are now equipping decoy cars with GPS devices to track and capture car thieves. Detectives installed a hidden GPS device in murder suspect Scott Peterson's car, and they knew whenever he drove past Amber Frey's house, knew he was at the bay to observe the body searches, and knew that he drove south toward Mexico, so they arrested him in San Diego.

13. Construction executives track stolen equipment. Between $300 million and $1 billion dollars worth of heavy equipment is stolen every year. Only 10 % is ever recovered. With a hidden GPS, the police can retrieve it, if loss of the equipment is reported before it reaches a border or a port.
14. Aircraft passengers can follow their locations. Systems are now being installed to permit passenger GPS to be used even when landing and taking off.
15. "Geocoaching" and "orienteering" hobbyists. Objects are first hidden by someone; other geosearchers then try to find them using hand-held GPS receivers.
16. Hikers and nature walkers use GPS to avoid getting lost or to help rescuers find them.

The GPS has 24 satellites. There are six orbits with four satellites in each orbit. They circle the Earth twice every day at 12,600 miles altitude. There are also three spare satellites up there, for use in case of failures.

Data received from the satellites can be merely stored at the tracking receiver or be retransmitted at regular intervals to a central location or to a computer by two-way radio or communications satellite.

The accuracy depends on several factors, including the receiver quality, but it is usually about 50 feet. The best achievable accuracy of a standard GPS receiver is about ten feet, and special systems can be accurate to one foot! Clocks on the satellites are compensated before they are launched for the difference in how fast the clocks will run due to the theory of relativity at the satellites' high speeds.

---

## Another Angle on Sunsets

At sunset, when the bottom edge of the Sun appears to just touch the horizon, the Sun is actually entirely below the horizon, with its top edge geometrically at the horizon! Refraction in the atmosphere causes the "rays" from the Sun to bend downward by about 1/2 degree, which happens to be the size of the Sun as seen from Earth. As a result, the entire Sun is still in sight even after it is entirely below the horizon.

Refraction also makes the bottom of the Sun look flattened at sunset. That's because the atmosphere is denser at lower altitudes. Light from the top edge of the Sun is refracted only slightly at sunset, and it curves downward only a little, but light from the bottom edge is refracted a lot more, so it has great downward curvature. Sighting along that downward curvature makes the bottom of the Sun look higher than it is, relative to the top of the Sun.

Then why doesn't the horizon itself have the same refraction as the Sun, so that the Sun would appear to be in the right place? The optical path of the horizon to the eye doesn't experience any change of refractive index of the media it traverses. Its entire path is along the surface of the Earth, where the density of the air is uniform. Sunlight, on the other hand, comes from outer space into the Earth's atmosphere. Like a beam of light going from air into a glass of water, the Sun's "rays" change direction when they enter the atmosphere.

Incidentally a sunset looks exactly like a sunrise, except that in the morning the atmospheric conditions are often different than at night. The Moon is refracted just like the Sun.

The Sun's image appears to be about the same size every day, because the Earth's orbit is not very eccentric (0.0017). The Moon is a little different. Someone might say "The Moon is big tonight." Actually, although the Moon's apparent diameter fluctuates more than the Sun's, it is not much more. The Moon's distance from the Earth varies from 221,000 miles to 253,000 miles, which is 13.5%.

But the Moon also has phases, and a full Moon probably looks larger because it is so bright. Like the Sun, the Moon subtends about 1/2 degree at the eye on average, as is demonstrated during a total lunar eclipse of the Sun.

When there is a crescent Moon we can still see the dark part dimly. That's because sunlight reflected from the Earth to the Moon illuminates the dark part of the Moon, and some of that Earthlight is then reflected back to the Earth. The Moon reflects only 7% of the light that strikes it, while the Earth reflects five times that percentage. That's because the Earth has white clouds and the Moon doesn't.

---

## The International Space Station (ISS)

It's a long time to stay away from wife and children. Six months is the average time for an astronaut to spend in the International Space Station. There were three crew members at first, but now changed to six, and the tours of duty of the six crew members will overlap. The windows are covered during their sleep hours to simulate nighttime. Otherwise they would see sixteen sunrises every twenty-four hours. The crew wakes up at 6 AM and inspects the craft, after which they have breakfast, review the day's plans by radio with Mission Control, and start work about 8 AM.

They work six days a week and exercise several times a day. They do scientific experiments such as studying the effects of space environment on humans. The results are expected to make lengthy space flights and space colonization feasible some day.

The International Space Station is the only space station presently aloft. Unlike most satellites, a space station doesn't have major propulsion or landing facilities, so other vehicles have to take things up and build it in orbit.

ISS was launched on Nov. 20, 1998, and after ten years portions were still being assembled by supply "missions." With 18 missions as of March 2009, it was about 81% complete. Completion is expected in 2011. It has had people on board since November 2, 2000. It uses solar panels for energy.

ISS is in a low orbit of 190 nautical miles above the surface of the Earth, where the crew is protected from solar flares by the Earth's magnetic field. Traveling 17,000 miles per hour, the ISS orbits the Earth every hour and half. An observer on Earth can see it at night with the naked eye if the station itself happens to be in Sunlight.

The ISS is the largest space station so far. It is 240 feet long and weighs 611,000 lbs. Large-scale space cities are being talked about but cannot be built in today's political climate because the launches are too expensive.

The cost of the ISS is expected to be $35 billion to $100 billion, spread over a period of 30 years. ISS is a joint venture of fourteen nations, including the USA. George H. W. Bush and George W. Bush supported it, and President Obama is expected to support it.

---

## Meteors, Meteorites, Meteoroids, Asteroids and Comets

Meteors When a meteoroid enters the Earth's atmosphere and starts to burn and glow from friction it becomes visible as a "meteor." A meteor is a streak of light seen in the night sky when a meteoroid burns itself out. The best observations are at four AM local time. Under clear conditions the eye can see about ten per hour. Most visible ones have magnitudes in the range + 3.75 to + 0.75. A meteor shower is a number of meteors with approximately parallel trajectories.

Meteorites Meteor fragments that survive the atmospheric fall are called "meteorites." They are interplanetary objects that fall all the way to the Earth's surface. The first documented one was a stone that fell in Europe in 1492, but it wasn't until 1803 that the scientific community accepted meteorites as being extraterrestrial. Roughly six observed "falls" and ten much later "finds" are added to the list annually. About 3300 hit the Earth each year, but most land in oceans, deserts and other uninhabited regions.

Meteoroids Meteoroids are chunks of rock in space, which are smaller than a planet or asteroid but larger than a molecule. Most of the meteoroid cloud around the Sun has particles smaller than 0.001 gram. Meteoroids are usually produced by the decay of short-period comets and the collisions of asteroids.

Asteroids Very large chunks of rock in space, some as large as mountains, are called "asteroids." Thousands of asteroids orbit between Mars and Jupiter, and a few have orbits that bring them near the Earth.

Comets Comets are different from asteroids in that they contain significant amounts of ice that vaporizes and glows when exposed to the light and to charged particles that stream out from the Sun. Comets are much larger than meteors and don't have to enter the Earth's atmosphere to be visible. A meteor streaks across the sky in a few seconds or less, but comets are visible for weeks as they slowly move against the background stars.

## Asteroid Risks To Human Life

There is a possibility that an asteroid will wipe out humankind, as happened to the dinosaurs. But we can still sleep tonight. The dinosaurs didn't have computer-assisted tracking or guided missiles.

Definitions

Tunguska A remote area in Siberia where in 1908 an asteroid (or comet) exploded 3 miles above the Earth. It caused physical damage comparable to the bomb at Hiroshima although the object was only about 130 feet wide. Reindeer herders 50 miles away were knocked down.

Asteroid A rocky body without atmosphere that orbits the Sun but is too small to be a planet.

Asteroid belt A region between Mars and Jupiter where about 95% of known asteroids circulate. Some maverick asteroids come inward and cross the orbit of the Earth.

Large asteroids Among those asteroids that cross Earth's orbit, there seem to be only 1000 or 2000 that are a mile wide or more. The devastating Yucatan asteroid of 65 million years ago was 5 miles wide. A large one could hit the Earth tomorrow, but maybe not for a thousand years. We'll probably be able to defend ourselves better 20 or 30 years from now.

Small asteroids Asteroids that are less than 150 feet across are much more numerous. The orbits of a hundred thousand of them cross Earth's path, and although small they can be deadly.

Impact Damage depends on whether the object strikes Earth on land or ocean. An ocean impact is worse and 70 % of Earth's surface is ocean. A tsunami wave caused by an asteroid could be a hundred times taller than the recent Earthquake-caused tsunami that killed thousands of people. If it hits on land, an asteroid only 500 feet wide could destroy a large urban area. A mile wide-asteroid could immediately wipe out all life in a large part of the Earth near where it strikes, and severely disrupt the rest of the Earth. The dust it raises could block the Sun long enough to kill food crops and create worldwide famine.

Detection methods Telescopes on Earth can detect large asteroids but can't detect objects as small as the 130-foot Tunguska one until they are dangerously close.

Protection Earth can be protected from an approaching asteroid by either destroying or deflecting it.

Destruction This method requires breaking the asteroid into small enough pieces that it will burn up in our atmosphere. An intruding asteroid could be blown apart by a nuclear bomb.

Deflection This method requires that an approaching asteroid be deflected perpendicularly to its trajectory. The Earth's radius is 4000 miles, so the new path preferably misses the center of the Earth by 5000 miles or more. An asteroid could be deflected by exploding a nuclear bomb near it, by striking it with a missile of great momentum or by landing a thrusting device on it and pushing it away.

Deflection by thrusting The earlier a pulse of deflecting thrust is applied, the less thrust is required, because there is more time for the altered trajectory to diverge before it gets near the Earth. If the asteroid were pushed slowly and steadily for several months until it passes us, a greater thrusting acceleration would be required to ensure a safe clearance.

Forecasts Cosmologists think that a billion years from now the Earth's environment probably won't be habitable for humans because of solar system aging. Nostradamus, a sixteenth-century French physician and astrologer, prophesied that the world would end in the year 3797. The Bible also predicted an end, but didn't say when: "Heaven and Earth will pass away...But of that day and hour no one knows, not even the angels of heaven, nor the Son, but the Father alone." Mathew 24:35-36. Some astronomers think it is quite possible that an asteroid will destroy civilization within the next million years.

Risk The chances of our dying by asteroid are estimated to be the same as those of dying in a plane crash, i.e., about one in twenty thousand. There are other threats to worry about that are more likely.

---

## Will We Become Extinct Like Dinosaurs?

Scientists are pretty sure that a meteorite landed about 65 million years ago near the Yucatan peninsula, killing all the dinosaurs and much of the other life on Earth. It was a 6- to 10-mile-wide asteroid or comet, and is estimated to have had the power of 30 million megatons of TNT.

One megaton is the explosive yield of a large nuclear warhead, and this was 30 million times as destructive. Mankind's predecessors have been dominant for only about two million years so far, so we came along much later. When a large meteorite hits, it causes dust that blocks the Sun, as well as heat and seismic effects.

Meteorites often hit the Earth, and scientists know how to estimate the risk and the damage that would result from any size. Impacts of about 0.02 megaton of TNT occur every year. There appears to have been a big explosion above ground in 1908 at Tunguska, Siberia. It may have been from a meteorite 18 miles in diameter, and equivalent to a 1-megaton nuclear detonation. Events this large or larger are expected about every 500 years.

The global catastrophe threshold is 200,000 megatons; one that big, or bigger, is expected to happen only about every 100,000 years. A meteorite as destructive as the dinosaur one is expected to occur only once in 4 million years, but of course they occur at random times.

Meteorites are solid interplanetary bodies that do not burn up entirely when they zoom in through the Earth's atmosphere.

Types of meteorites are stony, iron, and stony-iron. Thus, most meteorites have the physical characteristics of asteroids, which are small bodies of uncollected planetary material that orbit the Sun. There are several thousand asteroids; 95% of them are in a belt between the orbits of Mars and Jupiter. The orbits of some of the asteroids cross the orbit of the Earth.

Ceres, the largest known asteroid, has a diameter of about 1000 km. Asteroids survive the plunge to ground because they are very dense. Most meteorites are too dense to have come from comet-related meteoroids.

Since 1970 a few dozen meteorites have been found that have Martian chemistry. They were probably knocked off Mars by asteroid impacts, and happened to sail into Earth's orbit. Antarctica is a good place to find them because it is 3000 miles wide, and meteorites land on the white surface.

The chances are small that a meteorite will devastate life on Earth within many thousands of years in the future. Scientists are working on ways to protect Earth from bodies from outer space. Astronomers and governments might be able to detect threats in advance, alert the public, and perhaps even divert an approaching body by striking it in flight with a missile sent from the Earth or by breaking it up with an explosion up there before it reaches us.

What should we do in advance to protect against meteorites? Should we have a checklist, a shelter, and provisions for food, water, clean air and first aid, and perhaps even a gun to protect us from our fellow human beings? The answer is that we shouldn't do much at all in advance to protect ourselves, because the probability is negligible that a meteorite would ever affect us. There are many other risks that deserve our attention because they are so much more likely to occur.

---

## Van Allen Belts

Van Allen belts are layers of charged particles above the Earth. The possibility that such particles were trapped up there was suggested by several theorists before it could be confirmed by space rockets. In 1958 Dr. James Van Allen of the University of Iowa instigated a project in which they put Geiger counters on the satellites Explorer 1 and Explorer 3, and discovered the belts, which are layers of energetic plasma surrounding the Earth. Plasma is a fully ionized gas. The belts were then mapped out by Sputnik 3 and several other space missions. Similar plasma belts have been found around other planets.

Earth has two such belts—one of small diameter and a concentric larger one, connected by an area of reduced density of particles, so they are sometimes referred to as a single belt.

Picture an apple with a banana skin wrapped around it twice at its middle. The first wrap represents the lower Van Allen belt, which, at the equator, extends 400 to 6,000 miles above Earth. The second wrap represents the outer belt, which runs from 12,000 to 40,000 miles altitude. The "banana skin" extends to about 65 degrees north and south latitude, and is curved to come close to the Earth. The layers change depending upon solar activity and solar wind.

James Van Allen was born in Iowa in 1914, and was valedictorian of his high school class. He graduated summa cum laude from college. Van Allen experimented with making measurements in space by sending a balloon to its maximum elevation while carrying a rocket, then firing the rocket to go higher. He obtained a PhD in nuclear physics from the University of Iowa.

Van Allen worked on various important scientific projects, then became head of the physics department at University of Iowa. He received many honors in his lifetime. Professor Van Allen died in 2006.

The belts consist mainly of high-energy electrons and protons. Solar wind is the cause of most of the charged particles. Plasma is produced by the action of the Sun's ultraviolet radiation (and X-rays) on molecules, etc in the upper atmosphere. Once formed, the energetic protons and electrons are trapped in the Van Allen belt by the magnetic field of the Earth (magnetosphere).

The Van Allen belts are dangerous radiation sources for equipment and humans who venture into them. The International Space Station circulates at a 190 nautical-mile altitude, which is a safe orbit below the lower Van Allen belt.

Dr. Van Allen well deserves to have the belts that he discovered named after him.

References

For "The Amazing Global Positioning System"

1. U.S. Navy GPS Operations; Rocky Mountain Tracking, Inc.;
2. Wikipedia and other Internet sources on Van Allen belts.

For "The International Space Station"

3. "NASA International Space Station." www. NASA.gov/ mission_pages/ station/main/index.
4. http//en.wikipedia.org/wiki/International Space Station
5. Halverson, Todd, "International Space Station Approaches Key Turning Point." Space.com. 8/31/01
6. Trinidad, Katherine "NASA Assigns Space Station Crews, Updated Expedition Numbering" NASA.gov/home/ hqnews/2008

For "Will We Become Extinct Like Dinosaurs?"

7. "Earth Impact by an Asteroid." www.permanent.com/a-impact.htm 3/23/08
8. Science Screen Report. vol. 36, issue 6. www.nsf.gov/pa.
9. "Asteroid-impact Software Shows Where It Hurts. www. treehugger.com/files/2007/08/asteroid-impact.php

10. "Giant Asteroid Could Hit Earth in 2014." CNN.com/2003/tech/space/09/02/asteroid.
11. Daintith, John and Gould, William editors. Facts on File Dictionary of Astronomy 5th ed., 2006.

For "Meteors, Meteorites, Meteoroids, Asteroids and Comets"

12. Illingworth, Valerie, Editor.. 1994 Illingworth, Valerie, editor: "Facts on File Dictionary of Astronomy" Market House Books Ltd., 3d ed., 1994.
13. McGraw Hill Dictionary.

For "Van Allen Belts"

14. "Van Allen Belt." NASA. http://imagine.gsfc.nasa.gov, 2009
15. "Van Allen Radiation Belt." Wikipedia, 2009.
16. "James Van Allen." Wikipedia, 2009.
17 "The Solar Wind." http://hyperphysics.phy-astr,gsu.edu, 2005

# Chapter 2
# Seasons and Holidays

## Why Constellations Are Different in Summer

Jen, Student: Why are all the constellations different in summer than in winter? In summer I can see Cygnus and Sagittarius above the southern horizon, but in winter I can't find them anywhere. In winter I can see Gemini and Orion, but they are missing in summer.

Tom, Professor: They are there. The "celestial sphere" extends completely around us, and its variations are hardly noticeably at all.

Jen: Do you mean that Cygnus and Sagittarius are on the other side of the celestial sphere from Gemini and Orion?

Tom: Yes, and in its northern half.

Jen: But why can't we see all of them?

Tom: Jen, as a mature student you know that we can't see stars in the daytime, and in summer, the Sun is in the same direction from the Earth as Gemini is. If we tried to see Gemini in the summer we'd have to look toward the Sun during daylight. Six months later the Earth is on the opposite side of its orbit around the Sun. That's winter, and the Sun lies in the opposite direction from the Earth from Gemini, so we can see Gemini after dark.

Jen: Oh! Does the fact that the Earth's north pole is tilted toward the Sun in June have anything to do with it? In June our southern horizon is farther north on the celestial sphere by twice 23½ degrees, or 47 degrees, compared with December.

Tom: Yes our horizon is tilted because the Earth's axis of rotation is tilted, and you can see farther south on the celestial sphere in winter because of that.

Jen: What about the "summer triangle of stars?"

Tom: The summer triangle has three very bright stars–Vega, Altair and Deneb, which are the first stars visible in northern summer evenings. They can often be seen even from big cities. They are in the south, high above Sagittarius and Capricornus.

Jen: I want to see the summer triangle. What would it be the best time to look for it?

Tom: Well, the best time for viewing is the week between a last quarter Moon and a new Moon, on a night when there has been a deep blue sky before Sunset, which indicates that there isn't much haze. A park far from the city is best.

Jen: Could you help me find the summer triangle?" And what else might we see?

Tom: Some meteors, maybe. And we'd see the Milky Way for sure. It looks like a glowing band, because it has billions of stars.

Jen: We shouldn't need coats.

Tom: No, but you can't see any stars until well after 9 PM in summer, and they'll put us out of the park at 11 PM.

Jen: When should we go?

Tom: I'll call you.

Jen: I'm in the book.

# Legends about Summer Solstice

In Santa Barbara, California they put on a great parade with costumes and music to celebrate the June solstice. They enjoy hanging out with others at an outing to observe this tradition, seriously or playfully. People have observed the annual change in the Sun's elevation for thousands of years, and have created many legends to explain it, complete with rituals. Nowadays many groups meet at solstice time to socialize and conduct ceremonies.

Neopagans, of whom Wiccans are the most popular, have reconstructed ancient Pagan religions, so they observe "Midsummer" at solstice time. Some sing, chant, dance to drums and tell stories around a bonfire until dawn to welcome the Sun on June 21. The Goddess Mother Earth and the God Sun King are believed to have married in May. The Druids believed it was bad luck to compete with the Gods, so June traditionally became a popular month for weddings.

The Japanese celebrate the start of the summer season in June and the December start of winter with "Setsubun." Some Christian cultures celebrate the estimated birth date of St. John the Baptist with a feast from June 23 to June 25.

Our Earth spins on an axis that isn't perpendicular to the plane of its annual orbit. It is cocked sideways 23½ degrees. Earth is a gyroscope; it stays at that angle, pointing toward the North Star, all year. In June the northern axis is tilted toward the Sun, and at the other side of its orbit, in December, it slants away from the Sun.

In the Northern hemisphere, the Sun is up longest and highest on June 21. It is hotter in Northern areas in June than December because the Sun's rays strike closer to the vertical in June. At the equator, daylight is about the same duration all days of the year--about 12 hours, but you'd need a diagram to believe that.

The Earth's orbit is not a circle; it is an ellipse, with the Sun slightly off center at one focus. Earth is 91.4 million miles from the Sun when it is closest (perihelion), in January. When it is farthest, in July, it is 94.5 million miles away. That 3% change of distance makes the temperatures vary on Earth, but not nearly as much as the tilt of the axis does. Our surfaces take a while to absorb the heat and to cool off, so the seasonal temperatures don't peak until after the solstices. The June solstice is usually considered the first day of summer, although some call it midsummer.

It is hard to tell when the solstices occur because the rate of change of the Sun's noontime elevation is flat at those times. Centuries ago people measured the places on the horizon at which the Sun rose and set, about ten days before and after the solstices, when the changes are more rapid, then interpolated to find the solstices. Modern instruments do better.

The summer solstice has many names, mostly from ancient times: Alban Heflin, All-couples Day, Feast of Epona, Johanistag, Litha, etc. Some folks dismiss the ceremonies of other religions as being led by Satan. Others simply attend the solstice events and join in the fun.

---

## Six Definitions of Summer Solstice

We know what the summer solstice is, but how is it defined? Let me count the ways.

In the northern hemisphere:

1. The day of the year having the longest sunlight.

2. The instant in the year when the Earth is so located in its orbit that the inclination of its polar axis is toward the Sun (north pole toward the Sun).
3. The time of the Sun's passing a solstice, which occurs on June 21 or 22 to begin summer in the northern hemisphere and on December 21 or 22 to begin summer in the southern hemisphere. Because of leap years, the dates vary.
4. The summer solstice in the northern hemisphere occurs when the Sun is in the zenith at the Tropic of Cancer.
5. The time of year when the Sun is at its greatest distance north of the celestial equator. (The celestial equator where the extension of the Earth's equatorial plane cuts the celestial sphere.)
6. A point that lies on the ecliptic midway between the vernal and autumnal equinoxes and at which the Sun, in its apparent annual motion, is 23-1/2 degrees north of the celestial equator. Earth's north polar axis is toward the Sun.

---

## The Cold Winter Solstice

Some folks believe that Christmas, Hanukkah, and some other religious holidays are celebrated near the winter solstice because things start looking up right after that. You probably already know the astronomical explanation. The winter solstice is the location in the Earth's orbit at which, in the northern hemisphere, the inclination of the Earth's polar axis is directly away from the Sun. It is a point on the ecliptic at which the

Sun, in its apparent annual motion, is at its greatest angular distance (23.4 degrees) south of the equator.

The winter solstice is also defined as the instant at which the Sun reaches this point. It is the beginning of winter–the day with the fewest daylight hours and maximum hours of darkness. For those of us in the northern hemisphere, the Sun is then at its lowest point in the sky for the whole year. It happens about December 21.

Of course the other solstice (summer) is on June 21. The summer and winter solstices, called "solstitial points," are midway between the vernal and autumnal equinox points. Because of leap years the dates vary a little. At the winter solstice, for mid-northern latitudes, the Sun rises at a point on the horizon that is the farthest south of east for the year. For a few days afterward, the place on the horizon at which the Sun appears to rise stands still, then starts moving northward. And a few months later, happy days are here again (unless you're a skier)!

## Increasing Depression, Then a Glimmer of Light

"Doctor Reynolds, I think I'm simply depressed." "I understand, Nancy.

In winter we're cold, less active and we feel tired. Lack of sunlight increases the secretion of melatonin and induces melancholy and longer sleep. Exercise and bright lights can help by decreasing melatonin and increasing serotonin. It's a good thing we have Christmas and other traditional religious holidays to look forward to. We feast, dance, decorate a green tree with brightly colored lights and socialize with friends and relatives."

The December solstice does give us hope for brighter days ahead. While the Earth is spinning daily on its axis and going around the Sun once a year, the axis of the spinning isn't at a right angle to the plane of its orbit (the "ecliptic"). The spin axis is tilted 23.4 degrees from perpendicular.

Centuries ago in winter people had to live off their stored food and any animals they could capture. At more than 40 degrees north or south latitude some people did not survive. Understandably, early people were very happy to see the days start to lengthen again at the December solstice.

The spin axis of the Earth points close to the North Star throughout the entire year, because the Earth is like a big gyroscope. For half a year the northern hemisphere tips toward the Sun, from about March 20 to September 22. About June 21st the Sun gets as far north as 23.4 degrees, to the tropic of Cancer.

In the other half year the southern hemisphere tips toward the Sun, tipping farthest at 23.4 degrees to the Tropic of Capricorn, about December 21st. The exact calendar times and dates have a step change every leap year of the Gregorian calendar and they creep toward the step changes during the four years between. On the old Julian calendar the December solstice occurred on the 25th.

Christmas or Christ's Mass was at first banned by the Catholic Church because it originated in a third century pagan ritual, in the Roman Empire. It celebrated a Sun god who returned undefeated, and was coming back at the solstice. Pilgrims in Massachusetts tried unsuccessfully to ban Christmas altogether because of its heathen origins. It was many centuries before Christmas was finally accepted as a Christian holiday almost everywhere. Strange to say, Easter also started as a pagan ceremony.

No places north of the Arctic Circle (latitude 66.6 degrees) see the Sun at all on December 21 except for some refraction from below the horizon. All points south of the Antarctic Circle see the Sun above the horizon 24 hours that day.

Traditional solstice-related rituals of thirty two religions around the world, past and present, are described in Wikipedia.org under "Winter Solstice." Most religions have festivities and ceremonies when darkening days to change to lighter ones.

---

## Glorious Springtime

The many definitions of spring suggest different strokes for different folks.

For a Poet:

The time hath laid his mantle by
Of wind and rain and icy chill,
And dons a rich embroidery
Of sunlight poured on lake and hill.
Written by French officer Charles of Orleans
while he was a prisoner of war in England.

For an Engineer: An elastic device such as a coil of wire, that regains its original shape after being compressed or extended.

Naturalist: A small stream of water flowing naturally from the Earth.

Gardener: The season when it gets warmer and plants revive.

Student: Break Time. Whew! Let's party!

Criminal: To cause to release from incarceration.

Homemaker: Spring-cleaning time.

Bar habitué: "Spring chicken" = young person.

Diner: To offer to pay for lunch.

Astronomer: In the northern hemisphere, spring is the season extending from the vernal equinox to the summer solstice. (Surprisingly, the term "vernal equinox" is an informal term for the "dynamical equinox.")

The Sun crosses from south to north of the equator at the dynamical equinox. This northward "crossing" occurs about March 21; the hours of daylight and darkness are then equal. The vernal equinox presently lies in the constellation Pisces, and it precesses westward about fifty seconds of arc per year.

The dynamical or vernal equinox is the zero point for both the equatorial and ecliptic coordinate systems, although in star catalogs a "catalog equinox" is now used. A catalog equinox is the origin (zero point) of right ascension for the catalog, which is fixed for that edition and is a very close approximation to the dynamical equinox.

As we all know, the end of spring (summer solstice), occurs about June 21, when the Sun reaches as far as 23.4 degrees north of the celestial equator.

---

## Vernal Equinox

Celeste (mature student): I've heard about the equinoxes all my life, but what are they anyway?

Prof. Knight: Visualizing them is a mental exercise in solid geometry. Equinoxes are the two points at which the ecliptic intersects the celestial equator.

Celeste: Thanks, but that doesn't help me much, because I'm not sure what the ecliptic is.

Prof. Knight: Well, the Earth goes around the Sun of course, and the ecliptic is just the plane of the Earth's orbit.

Celeste: And what is the celestial equator, which you said intersects the ecliptic?

Prof. Knight: The celestial equator is an outward extension of the plane of the Earth's equator. It's perpendicular to the axis of the Earth. It serves as a reference plane for the right ascension and declination.

Celeste: Yes—I've also heard about right ascension and declination. They are used to give positions of stars aren't they? But what are they?

Prof. Knight: Right ascension and declination are coordinates of the "equatorial coordinate system." They are just extensions to the sky of the familiar latitude and longitude (respectively) that we use to specify positions on Earth.

Celeste: Oh! And what does the vernal equinox have to do them?

Prof. Knight: The position of the Sun at the vernal equinox has arbitrarily been taken to be the zero point of the equatorial coordinate system. By the way, even though the right ascension and declination are angles, they are expressed in hours, minutes and seconds instead of in degrees. An hour of declination is 15 degrees (which happens to be about how far the Earth rotates in an hour).

Celeste: Is the vernal equinox at the same place on the ecliptic every year?

Prof. Knight: No—but it hardly moves at all. Because of precession of the 23.4 degree tilt of the Earth's axis, the equinoxes move westward around the ecliptic through the constellations, completing a revolution each 25,800 years.

In the last 5000 years the vernal equinox has moved through two zodiacal constellations, from Taurus through Aries and into Pisces.

Celeste: Doesn't the vernal equinox usually occur about March 21st?

Prof. Knight: Yes. That's when the Sun appears to cross from south of the equator to north. On that day the Sun is neither north nor south of the equator, but over it. Incidentally, the vernal equinox is also called the spring equinox or dynamical equinox.

Celeste: I know that the autumnal equinox is about September 23d.

Prof. Knight: Yes. And on the days of the equinoxes the hours of daylight and darkness are equal. Only at the equinoxes does the Sun rise due east on the horizon and set due west. The locations on the horizon of the sunrise and sunset are changing the most rapidly on an equinox day.

Celeste: Are solstices part of this mental exercise in solid geometry?

Prof. Knight: Solstices are when the Sun is at its greatest angular distance (23.4 degrees) north or south of the plane of Earth's equator. They occur in June and December.

Celeste: Gosh, Prof. Knight—you sure know a lot. Thanks for staying after class to explain all this!

---

## Easter–Religious and Secular

"In your Easter bonnet,
With all the frills upon it
You'll be the grandest lady,
In the Easter parade..."

It's that time of year again. Special church services, Easter baskets, Easter egg hunts, live chicks and chocolate bunnies.

"Honey, I want a new dress for Easter." "You already have two dresses." "Something pastel--a hat would be nice too."

The Bible tells us that Jesus rode into Jerusalem with his disciples on what is now called Palm Sunday. On Maundy Thursday they had the Last Supper. Judas betrayed Jesus, and Pontius Pilate had Jesus crucified on the day many now commemorate as Good Friday. On Easter Sunday He arose from his tomb and briefly rejoined His disciples, then ascended to heaven.

The churches determine the dates for Easter Sundays; the ecclesiastical Easter date for each year is laid out on a list. It is the Sunday following the Easter ecclesiastical full Moon date for the year. That full Moon date is always the first ecclesiastical full Moon after the vernal equinox. Seventeen centuries ago March 20 was the equinox date. Ecclesiastical full Moon dates, which are pre-defined, are only approximate, and not always the same dates as the actual astronomical full Moon of modern times. Considering they were forecast about 326 AD the forecasters did a good job, but the dates may differ by as much as two days.

Later, the ecclesiastical rules decreed that the vernal equinox is always on March 21, and that date is now used to compute the Easter date. The actual astronomical vernal equinox, which isn't used to compute the Easter date, varies slightly from year to year on the modern secular calendar.

In 1582 Pope Gregory XIII of the Roman Catholic Church changed from the Julian to the Gregorian calendar, and that affected the dates of Easter Sundays. The Gregorian calendar, which introduced leap years, is very good, and was adopted throughout Western Europe gradually over the next few centuries. Some Eastern Christian churches still determine Easter dates by the Julian calendar, so the dates are different in different religions. Other people believe that Easter is based on the Hebrew Passover, and respectfully state that it commemorates the perfect Passover sacrifice.

A popular simplified version of the rule for the date of Easter in the USA is the first Sunday following the first full Moon after the vernal equinox. Usually this simplified version yields the same date as the official church-set date. Around the world, Easter in each time zone starts when the designated day starts for that time zone.

The Easter Sunday of the western religions can occur any time from March 22 to April 25 inclusive. About half of Easter Sundays occur on the same date 11 years later. (Trivia: The days on which Easter occurs repeat with a period of 1,900,000 years.)

In 2008 Easter came early because March 21 was followed the next day (Saturday), by an ecclesiastical full Moon. Sunday March 23 was Easter Sunday. Holy Week started on Palm Sunday, March 16. St. Patrick's Day, March 17, fell in Holy Week in 2008. In 2010 Easter Sunday was on April 4.

Because it commemorates the resurrection, Easter is a time of rejoicing. It is also a time for getting dressed up.

"Oh, I could write a sonnet,
About your Easter bonnet
And of the girl I'm taking,
To the Easter parade."

# Calendars–Solar, Lunar and Lunisolar

This is not just about your Day-Timer. A calendar is a system for keeping track of time, in which days are grouped into units, for regulating civil matters, religious observances, and scientific work. Calendars are based on the motion of the Sun, the Moon, or both, so they approximately fit either a solar year a lunar year. Today's calendars include the Gregorian (most popular), and the Moslem, Jewish, and Chinese

Solar Calendar: The Gregorian calendar is solar. Its average year is 365.2422 days. Pope Gregory XIII introduced it in 1582 to improve the Julian calendar that was established in 46 BCE by Julius Caesar. (The Gregorian calendar was in committee for 300 years.) The Gregorian calendar touched up the simple Julian calendar rule of having a leap year every four years, by skipping leap year in each "century year" (unless the century year was divisible by 400). Thus, 1600 and 2000 were regular leap years, but 1700, 1800, and 1900 were not leap years. There is still a small discrepancy between the Gregorian year and the tropical year (Sun year), but it is only about three days per 10,000 years.

Lunar Calendar: The lunar (synodic) month is the time between two successive new Moons. The average lunar month is 29.5306 days long. A lunar year is actually 354.3672 days. A lunar calendar has 12 months, each month having either 29 or 30 days, so it has a year of 354 days, with a leap year of 355 days.

Lunisolar Calendar: A lunisolar calendar is a lunar calendar that is brought into step with the solar or seasonal calendar by the addition of a 13th leap month. To keep festivals in season and bridge the 11-day gap between the lunar and solar years, the Jewish calendar adds a whole month 7 times in a 19-year cycle.

Easter: Easter is a mixed solar calendar and lunar-calendar event. It is always the first Sunday after the full Moon on or next after March 21, or one week later if the full Moon falls on Sunday. If you forget, no problem; just ask a candy company, florist, or church

References:
For "Vernal Equinox"
1. Zeilik, Michael: "Astronomy," John Wiley & Sons, Inc., New York. 7th ed., 1994.
2. Chaisson, Eric and McMillan, Steve: "Astronomy Today," Prentice-Hall, 3d ed. 1993.
3. Illingworth, Valerie, Editor. "Facts on File Dictionary of Astronomy" Market House Books Ltd., 3d ed., 1994.
For "Easter"
4. "The Date of Easter." http://aa.usno.navy.mil/faq/docs/easter.php
5. "Dates of Ash Wednesday and Easter Sunday." navy, loc. cit.
6. "Easter Dating." http://users.sa.chariot.net.au/~gmarts/easter.htm
7. "Easter." httm://en.wikipedia.org/wiki/easter
8. "Frequency of the Date of Easter."http://www.smart.net/~mmontes/freq3/html
9. "Easter." Eric Weisstein's World of Astronomy. http://scienceworld.wolfram.com
For "Increasing Darkness, Then Slowly a Glimmer of Light"
10. http://en.wikipedia.org/wiki.winter_solstice
11. www.religioustolerance.org/winter_solstice.htm
For "Legends about Summer Solstice"

12. "Popular Pagan Holidays" www.witchvox.com
13. "The 2009 Santa Barbara Summer Solstice Celebration."www.solsticeparade.com
14. "The Annual Summer Solstice Folk Music, Dance and Storytelling Festival" www.ctmsfolkmusic.org
15. "Celebrating the Seasons" www.circlesanctuary.org
16. "Summer Solstice Celebrations" www.religioustolerance.org
17. "Holidays and Observances" www.chiff.com/a/summer_solstice
18. "Summer Solstice" www.bbc.co.uk/religion/paganism/holydays

# Chapter 3
# Big Bang

## Hubble's Expansion Discovery

The radial velocity of a galaxy outside our Local Group is proportional to its distance from us. That's all Hubble was saying. A Monday morning quarterback might think, "I could have discovered that."

If a crowd of people on bicycles are together in the middle of a big field, and they all start riding in different directions at the same time, the fastest ones will be the farthest away after one minute. By seeing how far away a rider is you can tell how fast he has been going. It's the same with our universe.

In 1912 American astronomer Vesto Slipher discovered that almost every spiral galaxy had a redshifted spectrum, so it must be receding from us. Edwin Hubble was also an American (1889-1953). In the 1920's Hubble plotted a scatter diagram of velocity versus distance for remote galaxies, and drew a straight line through the points. It shows that the recession velocities of extra-galactic nebulae increase in direct proportion to their distance. Hubble's law would be explained by a uniformly expanding isotropic universe (Big Bang theory).

If you assume that the velocities of those galaxies have been constant, the slope of Hubble's line is a constant of proportionality for converting from distance to radial speed. You multiply distance by the "Hubble constant" to get radial velocity. The Hubble constant is not known very accurately yet, but is believed in year 2010 to be about 20 km per second per million light years.

Assume that everything started out at the same point. Then the age of the universe can easily be calculated by dividing the distance to a galaxy by its velocity. This means that the time of the Big Bang is simply the reciprocal of Hubble's constant. When you pay attention to the units, it turns out to have been about 14 billion years ago.

You might think that you have to be at the starting point to apply Hubble's law. Actually it doesn't matter which of the moving galaxies the observer is on. An observer on any galaxy would see the expansion. However, you should remember that the redshift as seen from Earth gives only the radial component of velocity.

NASA named a space telescope after Hubble and put it in orbit in 1990, launching it from the space shuttle. lt is 2.4 meters diameter and gets its power from solar arrays. The optical system initially had spherical aberration because the mirror was too shallow by 2 micrometers, but they repaired it three years later with small corrective lenses.

## From a Big Bang To a Little One

Human beings are aggregations of the elements carbon, hydrogen, oxygen, etc. There was a Big Bang 13.7 billion years ago, after which these elements evolved along with the rest of the present universe. Regardless of how we humans were created, this human aggregation of elements is now going to write the first page of the history of the Big Bang.

Scientists built the Large Hadron Collider (LHC) on the border between Switzerland and France, where they will duplicate events shortly after the Big Bang, but without the explosion. The LHC is a donut-shaped racetrack 17 miles in diameter and 150 to 500 feet underground. The tunnel is 12 feet wide and lined with concrete. About 5000 scientists and helpers are working on the project, and thousands more have contributed. Construction has gone on since about 1999.

What they are trying to do is confirm or modify the theoretical description of particle physics called the "Standard Model." They'd like to verify that the theoretical "Higgs boson" exists, which is thought to imbue other particles with properties such as mass, and may therefore account for dark matter in the universe. Also, <u>anti</u>-matter is presently unaccounted for; the amount was theoretically equal to the amount of <u>matter</u> created in the Big Bang, and these experiments may find out what happened to all that anti-matter.

Needle-shaped bunches of protons or of heavy ions of lead are going to be accelerated clockwise and others counterclockwise in a different pipe, to collide head-on at four points where the beams cross.

The project is gargantuan. More than 1600 superconducting magnets, each weighing thousands of tons, will keep the beams on track. The magnets are cryogenically cooled to 1.9 degrees Kelvin by a hundred tons of liquid helium.

The groups of ions that LHC will smash together will be accelerated in successive stages in several pre-accelerators. Particle detectors of several kinds are located at the four sites of collision crashes. They will detect some of the thousands of particles, traveling at almost the speed of light, produced by each collision. The tests will be done in stages of increasing complexity.

Physicists hope to rewrite our knowledge of physics by resolving the composition of energy and matter. The smashups should produce something called hadrons. Hadron are subatomic particles composed of "quarks". There are six kinds of quarks, which are particles having electric charges about one-third or two-thirds that of the electron. They may be responsible for the "strong interaction" which causes protons and neutrons to bind together in the atomic nucleus. Protons and neutrons are well-known hadrons.

The European Organization for Nuclear Research (CERN) built the LHC, whose total cost is about $8 billion (U.S.). There has been some criticism because the money could have been spent to reduce climate change, or poverty in Africa. Also, there was some concern and even court action about the risk of explosions threatened by the particle collisions, but scientists agree it is negligible.

On September 10, 2008 the first particle beams circulated successfully. Later, a defective connection between two magnets caused a shutdown, and a ton of liquid helium leaked out. A winter maintenance break was scheduled anyway, so operation of the LHC was not be attempted again until spring of 2009. An ambitious test and equipment upgrade is planned for the year 2019.

People are working very hard, and probably successfully, to ferret out the ancient history of our universe.

# Elusive Higgs Boson

The Higgs boson is going to be in the news a lot for the next few years, partly because of intense competition between two physics laboratories to verify that this particle exists. Scientists at the Large Hadron Collider at CERN laboratory on the Swiss-French border and at the Fermilab Tevatron Acelerator near Chicago are striving to detect Higgs bosons. So far there is no direct experimental evidence, although people have searched for over a decade.

What is a Higgs boson, and what does it have to do with astronomy? Subatomic particles such as electrons, protons and neutrons may have received their mass from Higgs bosons in the first few trillions of a second of the Big Bang. Theory says the transfer occurred by a process dubbed the "Higgs mechanism," but particle physicists are a skeptical lot; they want evidence.

Writers have nicknamed the Higgs boson the God Particle because it may have created the whole mass-energy universe we now have. Higgs boson is the only particle of the "Standard Model" (the textbook theory of quantum physics) that hasn't been detected yet.

The Standard Model is a possibly incomplete theoretical framework for describing the origin of the universe, the processes in the interior of the Sun, or the interactions of elementary particles. If a Higgs boson is found it will increase the credibility of the Standard Model, which unifies electromagnetism, the strong force, and the weak force. That includes everything except gravity, so it is a theory of almost everything.

The Higgs boson is the fundamental particle of the Higgs "field," which is a hypothetical quantum field that fills the universe and gives subatomic particles their mass. (All the fields have a particle associated with them because of field-particle duality.)

Peter Ware Higgs is a Scottish physicist born in 1929 who, among others, theorized in 1964 that there must be such a boson. Higgs recently asked his doctor to try to keep him alive until the theory is verified. Higgs bosons are expected to be verified by high-speed collisions between protons and protons, or by collisions between protons and anti-protons.

A mountain of data is produced when two such particles smash together at high speed. The two teams at the Fermilab, i.e., DZero and CDF (Collision Detector at Fermilab), recently combined their data to improve their chances of detecting a Higgs signal. The Fermilab accelerator is about 5 miles in circumference. A Higgs boson resulting from a collision would be so short-lived that the researchers have to look for the particles that it decays into. It might decay to two quarks or two leptons. Electrons, protons, and neutrons are believed to have leptons, quarks and bosons.

The CERN Lab's accelerator, funded by 60 nations including the USA, is an underground tunnel 17 miles in circumference that cost $6 billion to build. It was repaired after a 2008 accident. CERN has two teams, ATLAS and CMS, working with their two largest particle detectors. A collision of two particles there might produce two photons.

Competition will expedite the search projects. It is an exciting race, in which all humankind will be the winners if it gives us a better understanding of our cosmos.

# Cosmic Microwave Background Radiation–CMB

When you tune between TV stations, a few percent of the noise you see on the screen is from 14 billion years ago. That part is CMB radiation.

What is CMB? Cosmic Microwave Background. CMB is also called MBR (Microwave Background Radiation) and CMBR. It is electromagnetic radiation of wavelengths between 1 millimeter and 8 cm, with its spectral density peaking at about 160 Gigahertz. The power in our microwave ovens is in this frequency range. CMB is called cosmic because it started billions of years ago, long before stars were formed, and there is no other natural source of this radiation.

Where is it? It is ubiquitous. The whole universe is filled with these microwave photons.

Temperature and isotropy CMB is at a temperature of 2.7° K above absolute zero. It is almost isotropic–better than one part in a thousand. (Pronounce the adjective "isotropic," as isoTROPic; but pronounce the noun "isotropy" as iSOTtropy.)

Source? CMB is a remnant of the Big Bang.

How formed? Right after the Big Bang, particles and anti-particles continually annihilated each other and recombined in the intense heat. When cooling had progressed to where they could not annihilate each other anymore, there were excess photons still in existence. Back then, some of this electromagnetic energy was heat radiation, but its wavelength was stretched (redshifted) to microwave proportions by the expansion of the universe. The expansion of the universe "cools" radiation in inverse proportion to the fourth power of the universe's scale length. Expansion occurred during the last 14 billion years, while some of the radiation that is now arriving at Earth was on its way here.

How far back can we see? CMB radiation is the oldest radiation known. We can see back to about 350,000 years after the Big Bang, when the Universe was much smaller. Before that the primordial soup was opaque. But then neutral hydrogen atoms were formed, which interact only weakly with radiation, so the scene became transparent.

Discovery CMB was predicted theoretically by George Gamow in 1948 and by Alpher and Herman in 1950. Penzias and Wilson of Bell Labs first observed CMB in 1965, for which they received a Nobel prize in physics in 1978.

How is it detected? Metallic "dish" antennas focus the microwaves, and they are conducted by waveguides to bolometers or special transistors. The Earth's atmosphere interferes with Earth-based readings in the strongest CMB range, which is at a few millimeters of wavelength, but CMB is clearly measurable from NASA satellites such as COBE (1989-96) and WMAP (2001). The polarization was first measured with an interferometer at an Antarctic South pole station and reported in about 2002. As was predicted, it is weakly polarized.

Uses CMB is the most important evidence of the Big Bang theory, because its spectrum represents a thermal equilibrium condition like that of a black body radiator, which is reached only after a very long time. With extremely sensitive instruments cosmologists can detect small fluctuations in the temperature of the CMB, and from that information, learn more about the Big Bang and about the origins of galaxies and groups of galaxies.

# String Theory and Superstring Theory

String theory says that the "elementary particles" are not like points but are lines known as strings. Elementary particles are the basic building blocks of matter that don't contain any internal subcomponents. The strings are much too small to be examined or measured microscopically. They are able to form closed loops.

People have been trying for many years to develop a plausible theory that unifies the four fundamental forces of nature. These fundamental forces are believed to be the "strong interaction," "electromagnetism," the "weak interaction," and "gravitation." Up to now, gravitation is the only one that the Standard Model hasn't been able to explain.

When the idea of strings is applied to the problem of trying to unify the four fundamental forces, the theory is called "superstring theory." According to this theory, all four of the fundamental forces can be explained when you use a multidimensional framework called a superstring. Superstring theory says that all the fundamental forces became unified no later than a minuscule fraction of a second after the Big Bang.

A guitar string can vibrate in a second harmonic mode and other higher harmonic modes in addition to its fundamental low-frequency resonance. Similarly, a loop of string can vibrate in an infinite number of modes or frequencies. Moreover, ten or eleven dimensions are available, making for a lot more varieties. It is hard to picture so many dimensions, but that doesn't mean they don't exist, any more than the invisibility of television signals means they don't exist.

According to string theory, particles such as electrons aren't really particles at all–each is a vibrating loop of only one string. All of the strings are identical. Different frequencies and patterns of string vibration cause different masses and force charges, so they are perceived as different masses or forces. They seem different because they vibrate in different patterns.

One pattern of string vibration exactly matches the properties of the "graviton," so gravity can be explained by string theory and is included in it. Because string theory can explain all four of the fundamental forces of the physical universe, it is sometimes nicknamed the Theory of Everything.

References:

For "From One Big Bang to Another"

1. "Large Hadron Collider: The Discovery Machine:" Scientific American.
2. "CERN - The *Large Hadron Collider.*" *public.web.cern.ch/public/en/ LHC/LHC-en.html*
3. "Large Hadron Collider." www.boston.com/bigpicture/2008/08/*the_Large_Hadron_Collider.*
4. "Large Hadron Collider" http:\\en.wikipedia. org/wiki/Large Hadron Collider.
5. "*Large Hadron Collider*: Judge dismisses 'doomsday' lawsuit." Telegraph.co.uk

For "Elusive Higgs Boson"

6. Daintith, John and Gould, William editors. Facts on File Dictionary of Astronomy 5th ed., 2006..
7. "A Higgs Boson Without the Mess." Physical Review. 24 June, 2009.
8. "Higgs Boson: A Ghost in the Machine." TIME magazine, 9 April, 2008.

9. "Closing in on the Higgs Boson." Discover Magazine. 3 Oct, 2009.
10. "The Higgs Boson." Origins, CERN. Ideas. 2009
11. "Higgs Boson: Glimpses of the God Particle." New Scientist. Physics & Math. March 2, 2007.
12. "Narrowing in on the Higgs Boson." Ars Technica. March, 2009

"String Theory"
13. Hawking, Stephen "The Universe in a Nutshell" Bantam Books, N.Y., 2001
14. Greene, Brian "The Elegant Universe." Norton
15. Illingworth, Valerie, Editor.. 1994 Illingworth, Valerie, editor: "Facts on File Dictionary of Astronomy" Market House Books Ltd., 3d ed., 1994.
16. Internet: Numerous string theory sections

# Chapter 4
# Famous Telescopes

---

## Webb Telescope, Successor of the Hubble

This is the first of two sections on this subject.

A new space telescope is being designed to replace the aging Hubble telescope. It will orbit the Sun instead of the Earth. The James Webb Space Telescope (JWST), named for the second administrator of NASA, will work in the near-infrared region of the spectrum. It is expected to be launched by 2014 from a space center in French Guiana on the northern coast of South America. It will take JWST about 30 days to reach its orbit.

The telescope is 43 ft by 14 ft, and has 7 times more collecting area than the Hubble Space Telescope. The main mirror, at 21.3 feet diameter, is more than twice the diameter of the Hubble reflector. It was a challenge to fit the telescope and its shield into a rocket only 16 feet in diameter. It had to be packed in pieces that unfold in space.

The JWST must be kept very cold by blocking radiation from the Sun, Earth and Moon. An enormous shield 72 feet by 35 feet will be erected on one side of the telescope.

The telescope will be located at a Lagrange point. An 18th century French mathematician, Joseph Lagrange, predicted that there would be five points at which a third body would remain stationary with respect to two other orbiting bodies.

In this case, the first two bodies are the Sun and Earth, and the third is the JWST.

JWST will orbit around the Sun at the second Lagrange point, L2, which is in line with the Earth and Sun. L2 is a special point where the centripetal force from the curvature of the telescope's orbit is equal and opposite the pulls of gravity from the Sun and Earth. L2 is one million miles farther from the Sun than is the Earth. JWST will keep up with Earth as Earth orbits the Sun.

Normally an orbit greater than Earth's would require more than one year to encircle the Sun, but the gravitational pull keeps the JWST locked to the Earth in a one-year orbit. Since the Earth the Sun both stay in the same direction from the telescope, the shield need not be moved around much.

There is a small chance that repair missions can be accomplished during JWST's 5- to 10-year life by using an Orion manned spacecraft as part of NASA's Constellation program which is imminently in jeopardy of cancellation for lack of funding.

The JWST is a three-mirror anastigmat. It is somewhat similar to a Cassegrain telescope, in which incoming light strikes the main mirror, and the main mirror reflects the light back the way it came, to a small mirror in the center, which reflects it again through a small hole in the center of the main mirror. The JWST three-mirror type is more complicated, and has a flat final image plane and low aberration compared with a two-mirror telescope.

Part 2 tells more about the telescope, who is building it, the cost and what the mission is expected to discover about the universe.

# Webb Telescope, Orbiting the Sun

The James Webb Space Telescope (JWST) will replace the 1990 Hubble telescope. It will orbit the Sun instead of the Earth and garner information from the near-infrared region of the spectrum. It is expected to be launched by 2014. Designing the JWST is a daunting technical project.

The primary mirror is composed of 18 hexagonal segments. The mirror segments will unfold after the launch to form a 21.3-foot reflector. Miniature motors that are responsive to a waveform sensor will position the 18 segments.

JWST is powered by an array of solar panels and has an antenna for sending signals to Earth.

Its infrared sensors are extremely sensitive, because radiation from distant sources of 14 billion years ago are very weak. The infrared sensors are sensitive to radiation in the 0.6 to 28 micron (millionth of a meter) wavelength range, which is near infrared. Infrared is better for this mission than visible light because it can see through clouds of dust in the viewing path.

The sensors are so sensitive that infrared (heat) radiation from the telescope itself would create an intolerable background noise. The Sun/Earth shield lets the telescope cool to minus 387 degrees Fahrenheit (40 degrees Kelvin). One instrument on the telescope is further cooled down to minus 447 degrees F (7 degrees K) by a cryocooler.

Although the largest ground-based telescopes are larger, JWST will have much sharper images because there is no atmosphere between the stars and the telescope. The JWST can scan the universe at the beginning of time by examining infrared radiation that started traveling toward us from shortly after the Big Bang.

It will show the formation of stars, galaxies and planetary systems. JWST will also see very distant galaxies in the process of formation and will provide information about the distortion of light by dark matter, which is an important topic in astrophysics today. It may disclose when the chemical elements we now know were first formed.

Over 1000 people around the world are working to develop the JWST. The main contractors are in the USA–Northrop Grumman and Ball Aerospace . NASA is the leader of a coalition of 15 collaborating nations, the European Space Agency and the Canadian Space Agency. The life-cycle cost is estimated at $4.5 billion.

The James Webb Space Telescope is an enormous and courageous leap of science.

---

## Fantastic Liquid-mirror Telescopes

A liquid-mirror telescope uses a liquid as a reflecting surface. The mirror is spun so its liquid surface becomes a paraboloid. A detector, which is usually one or more charge-coupled devices (CCD), is mounted in the center above the mirror, just as in many conventional telescopes. The rim of a typical 4-meter mercury telescope spins about 3 miles per hour.

Isaac Newton proposed liquid-mirror telescopes, but electric motors with which to spin them smoothly had not been invented yet. Their main drawback is that they always look straight up at the zenith. If they are tilted the mercury spills out.

One might think such a telescope would be almost worthless, but not so. The Earth is rotating, so our zenith is continuously moving. A star that comes into the telescope's field of view crosses the zenith at 15 degrees per hour, which is 1/4 degree per minute.

Many scientific groups are working on these telescopes. An International Liquid-mirror Telescope 4 meters in diameter, funded by Belgium and Canada, is being built on a mountain top in India. It has a 1/2 degree field of view. An 8-meter telescope intended for a mountain in Chile will have 240 CCDs. Its field of view is **3** degrees, so a star passing through the center will be visible for 12 minutes.

Some tracking of stars is possible. If the detector is a CCD, it can be of the type which tracks a moving image across the CCD. A secondary mirror mounted above the mercury mirror can be warped mechanically to change direction slightly and can prolong the viewing time of a star by following it.

Research is underway to develop liquid-mirror telescopes that can be tilted. A group at Laval University in Quebec, Canada is developing reflecting liquids of high viscosity which can be tilted up to 30 degrees.

A liquid telescope costs only one or two percent as much as a glass one of the same size, and is cheaper to maintain. The liquid-mirror need not be cleaned, adjusted or aluminized. Conventional glass telescope mirrors are meticulously cast, ground and polished, then reflective aluminum is vaporized in a vacuum to cause an aluminum film 100 nanometers thick to be deposited on the glass.

A zenith-pointing telescope can be used as well as any other for some cosmological studies, because the universe is isotropic and homogeneous and they can study it in any direction. An 8-meter telescope looking straight up can detect many supernovae and see as many as a billion galaxies as the Earth turns. Gravitational lenses can also be discovered and analyzed with these telescopes. They can also be used to study the properties of dark energy. It is possible that liquid-mirror telescopes will replace many glass ones in the near future.

Mercury is the most popular liquid for liquid telescopes but gallium alloys and others are also suitable. The mercury layer is only 0.5 millimeter to 1 mm thick on the surface of the spinning bowl because thicker layers cause distortion.

The bowl is usually as close as possible to the shape that the mercury will assume at the planned rate of spinning. The bowl need not be perfect because small irregularities in its shape don't hurt much. The bowl has a stainless steel compressed-air bearing whose thin film of air has almost no friction. A three-point mount aligns the axis of rotation parallel to the gravitational field of the Earth.

When the mercury is spinning, two forces act on it–gravity and centrifugal force. Gravity pulls the mercury down and the inertia of centrifugal force pulls it radially sideways. A simple algebraic equation based on balancing the forces shows that the cross section is a parabola. If the rate of spinning is increased the parabola becomes steeper and has a shorter focal length. Focal length varies inversely as the square of the rotational RPM.

Homemade liquid-mirror telescopes have been made with gramophones, but they are dangerous because mercury is a poison.

Next month's article will describe a 66- to 328-foot liquid-mirror telescope proposed by NASA for installation on the Moon as early as 2020. The absence of an atmosphere and the telescope's great aperture will provide spectacular advantages. The telescope will see light that was emitted 13.5 billion years ago, shortly after the Big Bang.

# Liquid-Mirror Telescopes on the Moon

Courage, thy name is NASA. Against daunting odds, they are planning to build enormous liquid-mirror telescopes on the Moon.

NASA is shooting for the year 2020 for the first one, which is likely to be 66 feet in diameter. Later they want to build one 328 feet wide, which would be the largest optical telescope anywhere. Weak gravity on the Moon will be helpful in building a giant telescope.

Fundamentals of liquid-mirror telescopes are described in the preceding section. Liquid takes a parabolic shape when its cylindrical container is rotated. The liquid-mirror always points straight up, so as the Moon rotates, the telescope sweeps a strip of constant declination across the sky. The telescope will probably be located near the Moon's north or south pole, where it could observe the same area of the celestial sphere continuously for months.

E. F. Borra of the Laval University of Quebec said "If the telescope is located anywhere than exactly at the poles, with each rotation of ...(the) Moon it would scan a circular strip of sky. And the rotational axis of the Moon wobbles with a period of 18.6 years; so over a period of 18.6 years the telescope would actually look at a good-sized region of the sky." The Moon's axis is tilted about 6°.

Mercury can't be used for the mirror because it would be frozen solid at the -240° F temperatures of the Moon. Organic compounds called "ionic fluids" are liquid at rather low temperatures; consisting only of ions, they are very viscous and won't evaporate. Physicists are now trying to develop ionic liquids that would remain liquid at Moon temperatures.

To make a Moon telescope reflective, a thin layer of silver could be deposited on the ionic fluid while it is spinning. The silver could be less than 100 nanometers thick. It would solidify and lie on the liquid.

Electromagnetic waves are greatly absorbed and attenuated by the Earth's atmosphere (except for visible light and radio-frequency waves). On the other hand, a telescope on the Moon, where of course there is no atmosphere, would also see microwaves, X-rays, gamma rays, ultraviolet light and infrared light. Moreover, a Moon telescope would not have the atmospheric distortion (e.g., scintillation) that plagues the signals that get through to the Earth.

The infrared region is important because light from the most distant parts of the universe is redshifted very far. The telescope will likely be located in a permanently shadowed crater where the temperatures are always cryogenic, because low temperatures are beneficial for infrared astronomy. Power sources that absorb sunshine could be nearby outside the crater.

Detectors mounted above the mirror could comprise hundreds or even thousands of CCDs (charge-coupled devices). With the "drift scan technique" for electronically scanning CCDs, a star can be tracked as long as it is in the telescope's field of view; the signals can be time-integrated. Star tracks are likely to be slightly curved because of the Moon's motion, and CCD scanning is in a straight line, so there is some scanning distortion. Glass corrector lenses are available to correct the scanning distortion introduced by the drift scan technique.

The Moon has no air, so compressed air bearings of the type used on Earth aren't practical there. The spinning mirror might be supported by a magnetic bearing as is done with magnetically levitated trains, and stabilized by superconducting components.

The container of the liquid must be light and rigid. It would no doubt be folded up for transport to the Moon, then unfolded like an umbrella. The container will be parabolic in shape, and after deployment will be filled with the ionic liquid.

The reason the liquid-mirror takes a circular parabolic shape is this: Heuristically, think about an atom of liquid at the bottom of the layer of liquid, resting directly on the spinning container. The atom turns with the container because all the forces acting on it are balanced.

Assume the container is parabolic, so the place where that atom sits is slanted relative to the axis. The supporting upward force provided by the container acts along a tilted line perpendicular to the container at that spot. The supporting force has a vertical component that is equal and opposite to the pull of gravity. It also has a horizontal component, which is equal and opposite to the centrifugal force of rotational acceleration. With balanced forces, the atom rides around in the same spot on the container.

Since the centrifugal force varies as the square of the radial distance of an atom from the central axis, the mirror can be shown to be a circular paraboloid, even when there are more than one "layer" of atoms.

A Moon-based liquid telescope having a wide spectrum, little distortion and hundreds of times the aperture of earlier telescopes could probably see 13.5 billion years into the past, which is within a half-billion years after the Big Bang.

Great things are ahead from the intrepid scientists of NASA (assuming they receive the necessary funding). This Moon exploit could put Columbus' voyages in the shade.

# World's Largest Optical Telescope

Dr. Kosmo, astronomer: Being a science teacher, Stella, you will be interested in a 140-foot diameter telescope that is going to be erected in Chile or in the Canary Islands. It's called the European Extremely Large Telescope, or E-ELT. The largest optical telescope now in use is 33 feet.

Stella: Why are they going to build such a big one?.

Dr. Kosmo: The greater light-catchment area of a large telescope enables it to see weaker light and objects farther away. This one will gather fifteen times as much light as earlier telescopes.

Stella: Doesn't atmospheric distortion make the images fuzzy?

Dr. Kosmo: Yes. Scintillation occurs as the light passes through the air, like the shimmering above a hot road. The image in a telescope then moves around and gets smeared. In the case of a wide light source like a nearby planet, scintillation makes the outline and features of the image fuzzy.

Because of atmospheric distortion a 1-meter telescope doesn't resolve much better than a 10-cm telescope, but of course the 1-meter one can detect dimmer objects. Under the very best seeing conditions (which are rare), the atmosphere imposes a resolution limit of 0.35 arcsecond.

Stella: If the resolution is limited by Earth's atmosphere anyway, doesn't a 14-foot telescope have as good an image, at least of bright objects, as a 140-foot one does?

Dr. Kosmo: Usually it does, but this E-ELT 'scope will have "adaptive optics" that compensate for the turbulence in the atmosphere, so it will have beautifully sharp images. It is expected to have much better resolution than even the Hubble space telescope.

Stella: What will the E-ELT be used for?

Dr. Kosmo: It will enable detailed studies of black holes, the nature of dark matter and dark energy, and galaxies having redshifts as great as 7, that were formed soon after the Big Bang (stellar archeology),. It is expected to find sister planets in other solar systems that are not too hot or cold for habitation, and that may have water and plant or animal life. It can also study the atmospheres of extrasolar planets. More uses will probably be discovered when the E-ELT starts operating in 2017.

Stella: What kind of telescope is the E-ELT?

Dr. Kosmo: Well, here are some scientific details:

- E-ELT is a reflector-type 5-mirror anastigmat (one which can create an image almost free of chromatic aberration, coma and astigmatism).
- Three of the mirrors in the light path are conventional, and two are deformable mirrors for correcting for the atmosphere.
- The 140-foot mirror is composed of 906 hexagonal mirror sections, each 14.6 feet across. The sections are adjusted by computer-controlled actuators.
- Focal length is 1379 feet to 2758 feet (f/10 to f/20).
- The mounting is of the altitude/azimuth type.
- Angular resolution is 0.001 to 0.6 arcsecond, depending upon how the telescope is used.
- Diameter of the field of view is 9 arcminutes.
- It will be sensitive to light in the optical and near-infrared ranges, from blue atmospheric cut-off to mid-infrared.

Stella: That's fantastic! But it sounds pretty expensive.

Dr. Kosmo: The design phase will cost $85 million and construction will cost $1.5 billion. It is now in the detailed design phase. The European Southern Observatory is in charge of the E-ELT project, which is funded by fourteen European nations.

Stella: Why don't they just put another telescope in space orbit, where there are no atmospheric problems at all.

Dr. Kosmo: Yes, while an uncompensated Earth-based telescope's image of the large Andromeda Galaxy is very fuzzy, Hubble images show its features very clearly. Unfortunately, space telescopes are much more expensive than Earth telescopes and are harder to up-date and maintain.

Stella: What is in the future for ground-based telescopes?

Dr. Kosmo: Telescopes on Earth having adaptive optics may supersede telescopes orbiting in space because adaptive optics is becoming very sophisticated and space telescopes cost so much.

References:

For "Webb telescope, Orbiting the Sun"

1. "The James Webb Space Telescope." NASA. jwst.nasa.gov/comparison.
2. "JWST Overview." European Space Agency. Oct. 21, 2009. SC120370.
3. "James Webb Space Telescope." encarta.msn.com/encyclopedia.
4. "Meet the Supreme Space Telescope. " Softpedia.com/news.
5. "Comparison of On-Axis Three-Mirror-Anastigmat Telescopes." mlampton@berkeley.edu

For "Fantastic Liquid-mirror Telescopes"

6. "Liquid-mirrors, a Long Expected Reality." Softpedia.com. 2009.
7. "Liquid-mirror Telescopes." American Scientist.org/pub/2007/3.

8. "The 4-m International Liquid-mirror Telescope Project (ILMT)."adass.org/adass/proceedings/00/P3-13/

For "Liquid-mirror Telescopes on the Moon"

9. "Liquid-mirror Telescopes on the Moon." Science@NASA.gov. Oct.8, 2008.
10. "Liquid-mirror." Special Dictionary, Encyclopedia. Dec. 24, 2009.
11. "Liquid-Mirror Telescopes." American Scientist. March 2007.
12. "The LMT Didactical Experim." A.Surdej & A.Pospieszalka-Surdej. Undated.
13. "Liquid-mirror." Wikipedia.org.wiki. Nov. 26, 2009.

For "World's Largest Optical Telescope"

14. "World's Largest Telescope Will Search Heavens for Habitable Planets Like Earth." http://www.telegraph.co.uk/science/space/5104408
15. "European Extremely Large Telescope." http://www.encyclopedia.com/doc/1080.
**16.** "European Extremely Large Telescope." http:en.wikipedia.org/wiki

# Chapter 5
# Sun and Moon

---

## Personality Profile The Sun

- Popular "The Sun has got his hat on; he's coming out today," (British song). And we'll be glad to see him. A charismatic guy, the Sun has often been worshipped as a god, with names including Ra, Helios and Apollo. Norse mythology cast the Sun as a female named Sol (and the Moon as a male)-
- Beneficial Plants and animals couldn't live without this fellow.
- Huge He's 240,000 miles in diameter, which is small as stars go, but thirty times the diameter of Earth.
- Old Middle-aged, at 4.6 billion years.
- Stand-off-ish. He lives 93 million miles from the Earth
- Heavy 2.2 x $10^{27}$ tons. But slimming–he's losing mass at the rate of 4 million tons per second.
- Hot 15 million degrees Kelvin at the center and 5700 degrees Kelvin at the photospheric surface.

- Energetic The Sun generates power by a nuclear fusion process deep inside, and sprays a total flux of about 1370 watts per square meter onto the Earth.
- Colorful He gives us all the colors of visible light, along with infrared radiation and more. The solar spectrum peaks at green, and the human eye has evolved to be sensitive to the range of wavelengths at which the radiation is strongest. Our atmosphere reduces the intensity of the peak wavelength by about one third.
- Psychological Sunligbt makes us happy. Artificial light therapy is being prescribed by psychologists for people who are depressed in winter. Sunlight is even said to be an aphrodisiac!
- Gaseous Consists of 74% hydrogen, 25% helium, and some heavy elements.
- Rotating It lumbers around on an axis tilted 7.25 degrees from the ecliptic, with a period of 25 days at its equator and 34 days at the poles.
- Warming Of course the Sun creates global warming, by as much as 58 degrees Fahrenheit. Water vapor, carbon dioxide, methane and nitrous oxide in the atmosphere absorb radiation from the Sun, and warm the Earth directly This component of warming is misnamed "greenhouse effect," from an incorrect analogy. There is also a bona fide greenhouse-effect component of warming in which incoming radiation reaches the land and water and is transformed into heat. Reradiated infrared radiation can't get back out through the

atmospheric gasses, so the Earth is warmed the way a greenhouse is.

- Noisy Radio-frequency static from the Sun disrupts our communications. It is worst at times of high Sunspots, every 1l-years. (The cycle takes 22 years if you consider their reversal of magnetic polarity).
- Dangerous Kids may not know they can hurt their eyes by looking directly at the Sun without special dark filters. And Sunlight can cause skin cancer, so tanning spas are risky.
- Mortal Presently a yellow dwarf star, the Sun will become a red giant star in another 4 or 5 billion years, then collapse to be an extremely dense white dwarf.

---

## Life Story of Old Sol

A biography of the Sun, told through definitions.

1. Starbirth Stars are formed in molecular clouds, by gravitational contraction of the cloud material. About ten stars per year are being formed in the Milky Way. This was the birthing stage of our Sun, which occurred about five billion years ago.
2. Protostar. The protostar stage is the portion of a star's life before the Main Sequence (see below). This early formation interval, which corresponds to the infancy of a human life, lasted about 50 million years for the Sun.

3. Evolution of a Star. During a star's life it burns with thermonuclear reactions, which cause changes in its radius, luminosity, surface temperature, and chemical composition.
4. Herzsprung-Russell Diagram (H-R diagram). This is a scatter diagram of many stars. Each star's surface temperature is on the horizontal axis and its luminosity is on the vertical axis. A Herzsprung-Russell diagram shows a number of stars that are in different stages of their lifetimes. By studying the present characteristics of many stars that are similar except for their ages, astronomers can theorize about how the Sun developed and how it will die.
5. Solar Mass Stars. A star's longevity depends mainly upon its mass and chemistry, but also on some other factors. Solar Mass Stars are ones that are about as massive as the Sun. Alpha Centauri is an example. The mass of the Sun is arbitrarily assigned the value 1.0 in some literature.
6. Zero-Age Main Sequence Star (ZAMS star). When a protostar becomes mature enough to derive most of its energy from thermonuclear reactions instead of from gravitational attraction, it takes a ZAMS position on the Main Sequence chart (see below).
7. Main Sequence. The principal series of stars in the H-R diagram is called the Main Sequence. It is a string of those many stars on the diagram that are in the phase of converting hydrogen to helium in their cores, by thermonuclear reactions. The Sun doesn't cool during this time even though it is radiating tremendous energy into space, because it is obtaining the energy from thermonuclear reactions. This is the longest stage of

a star's active life, corresponding to the adult life of a human. The Sun will spend about 80% of its life in the Main Sequence.

8. Red Giant. A large star with high luminosity and low surface temperature. When the fuel for thermonuclear reactions runs out and the star is still hot, its surface temperature starts to decrease, and the radius expands tremendously. Our Sun will probable become a Red Giant later in its life, with a radius greater than the present distance to Mercury.
9. Expulsion of Outer Layers. The Sun will develop a strong outflow of mass from its surface, in gusts that rip off the envelope. The envelope moves outward in shells that continue to expand, leaving a hot core. This corresponds to the phase when a human first retires from work.
10. White Dwarf. The core that remains after expulsion of outer layers becomes a White Dwarf, which is composed mostly of carbon that is not hot enough to burn. All thermonuclear reactions have stopped, and there is slow cooling. When the Sun is a White Dwarf it will be in the phase of its life similar to a human's retirement.
11. Black Dwarf. This is the cold remains of a White Dwarf after all its thermal energy has been exhausted. The Sun will be a corpse at this phase of its ten billion year life.

# Orbits–Copernicus & Kepler

Orbits are crazy. Our Moon and the planets are always on their good behavior, but Halley's comet is really eccentric. It loops around the Sun like a road-rage driver. Orbits happen in astronomy when an object is moving in a gravitational field, such as the Moon traveling around the Earth or the Earth around the Sun.

Planets, moons, comets, binary stars and man-made satellites are all in orbit because they are in the gravitational field of another body nearby. Many great astronomers had theories about orbits.

About eighteen hundred years ago, Ptolemy proposed that the other planets and the stars were in spherical configurations centered on the Earth. Copernicus recognized in 1514 that the planets' orbits were centered on the Sun, not the Earth, but he still thought they were circular. Kepler used Brahe's empirical data in 1609 to conclude that orbits are elliptical, but incorrectly thought the planets might be held in their orbits by magnetism.

Newton calculated in 1680 that an inverse-square law of gravitational attraction between the Sun and the planets would explain the elliptical orbits that Kepler described. Einstein proposed in his general theory of relativity (1915) that gravitational and inertial forces are equivalent and that curvature of space is the physical basis for orbits.

Of course Copernicus was right when he said that the Earth revolves around the Sun, but his model (like earlier ones) still had circular orbits. Johannes Kepler changed that in 1609. For a single object in the gravitational field of a big body like the Sun, the orbit is a conic section, usually an ellipse or hyperbola. An ellipse is a partially flattened circle, like the Earth's circuit around the Sun. In a hyperbolic orbit the orbiting object escapes from the larger body, like the space probe Voyager 2 leaving the Earth.

Kepler's first law says that the orbital paths of the planets (for example, Mars and Saturn) are elliptical, with the Sun located near one focus of the ellipse. (The other focus is in empty space.)

Kepler's second law states that the speed of a planet varies in its orbit. lt moves faster at perigee (nearest) and slower at apogee. An imaginary line connecting the Sun to a planet (for example Jupiter) sweeps out equal areas of the ellipse in equal intervals of time.

Kepler's third law tells how a planet's period is related to the size of its orbit, larger orbits requiring longer to be traversed. The square of a planet's orbital period is proportional to the cube of its semi-major axis (half of the "long way across"). For example Saturn, which is almost 10 times as far from the Sun as the Earth, takes a lot more than 10 Earth years to go around the Sun; it takes nearly 30 Earth years.

Kepler was substantially right about orbits. lsaac Newton later refined the first and third of Kepler's laws, so all three laws apply to all bodies in Newtonian mechanics.

---

## Mercury as an Example of Orbits

Why don't planets just go in circles? What's their problem? Why are their orbits elliptical?

Nomenclature  We know that an ellipse, which looks like a flattened circle, is a two-dimensional planar curve. It's a conic section whose plane is slanted from the intersected cone. The <u>sum</u> of the distances from any position on the ellipse to two fixed points situated inside it, is equal, for all positions. The two fixed points are called the foci. Eccentricity is the distance between foci divided by the length of the major axis (for the Sun, from aphelion to perihelion). The eccentricity of an ellipse is less than one. A circle is a special case having zero eccentricity.

An orbit having an eccentricity greater than one can be hyperbolic, with the body's velocity exceeding the escape velocity. As it passes the Sun it changes direction and goes to outer space, never to be seen again.

Mechanics of an orbit In Newtonian mechanics, the two main forces acting on our planets are the centrifugal "force" and the gravitational attraction of the Sun. The faster a planet moves the greater is the centrifugal force trying to fling it away. Centripetal force (inward), which is the opposite of the reactionary centrifugal force, is merely Newton's familiar law that "force equals mass times acceleration." It is in a form that enables that law to apply to a mass moving on a curved path.

The eccentricity and other parameters of a planet's ellipse depend not only upon its mass but upon the initial conditions with which it entered that orbit, such as (a) distance from the Sun, (b) speed, and (c) direction of travel.

Mercury's eccentric orbit The planet Mercury is a good example. It has a rather eccentric orbit, with a perihelion of 28 million miles and aphelion of 43 million miles. Imagine a picture of an ellipse, with the Sun and the perihelion near the bottom of the page and the aphelion at the top.

Inertia tends to make Mercury go straight ahead, tangent to its orbit, but gravity makes it turn toward the Sun, which is at one focus. That change of direction is a form of acceleration.

As Mercury curves along its orbital path, the centrifugal reaction prevents it from spiraling into the Sun.

Mercury from perihelion to aphelion Let's follow Mercury, starting at a point in its orbit part way up the page on the right. Mercury's velocity is so great there that the pull of gravity can't curve the path fast enough to put the center of curvature at the Sun. (Gravity produces acceleration of an orbiting body, and can divide its effects between changing the orbital curvature and changing the orbital velocity, depending on the body's direction of motion.) At that orbital position a small portion of the force of gravitational attraction is apportioned to speed (orbital velocity) acceleration, which in this case is negative--a deceleration. Mercury's speed decreases while it heads upward on the page from perihelion to aphelion.

Mercury at aphelion When Mercury is at aphelion, at the top of the page, it is moving slowly enough for gravity to have time to create a sharper curvature. All of the gravitational acceleration is allocated to curvature and none to change of speed. At aphelion the curvature becomes so sharp that the radius of curvature does not even reach all the way down to the Sun.

Mercury from aphelion to perihelion As Mercury proceeds down the left side of the page, its direction points very slightly toward the Sun, so the gravitational force acting on it provides a small component of linear acceleration that speeds it up. It keeps going faster until it gets to the midpoint of the perihelion.

Mercury at perihelion At perihelion Mercury is so close to the Sun that, because gravity is so strong there, it again produces a sharp curvature despite the fact that Mercury is traveling very fast at that point. Altogether, the resulting path is an ellipse.

## Conclusion

Although the mass and speed of the orbiting body determine the size of its orbit, the eccentricity depends entirely on the initial conditions.

Mercury's orbit doesn't become circular because Mercury doesn't pass through any position corresponding to initial conditions of a circular orbit. At some positions in Mercury's orbit the curvature is perfect for a circular orbit, for example near a "northwest" position on the page. Nevertheless, the orbit doesn't convert to circular there because the center of curvature is not at the center of the Sun at that time. Elliptical orbits are stable and they don't change unless there is a force to change them.

That's why all planets don't simply sail around in circles.

Incidentally, the eccentricity of an elliptical orbit gradually decreases over many millions of years because of secondary effects, so orbits do slowly get closer to being circular

---

# Solar Wind–What Is It?

The Sun's corona is so hot and energetic that the Sun can't hold onto all of it. Solar wind is a flow of charged particles, mostly protons and electrons, flowing from the Sun's corona into space. The particles overcome the gravitational pull of the Sun, exceeding the escape velocity.

Solar wind causes the visible tail of a comet to be directed away from the Sun, even when the orbiting comet itself is moving away from the Sun. In the early 1600s Kepler guessed that comet tails were under pressure of sunlight, which is correct as far as the dust tails are concerned. Comets also have ion tails.

In 1943 Cuno Hoffmeister in Germany said that the Sun also puts out particles. In 1948 Eugene Parker of Chicago explained the outflow of particles while he was studying the Sun's corona.

Solar wind is a plasma, which is an ionized form of matter having equal positive and negative charges, and it carries with it some trapped magnetic fields. An electron traveling in a magnetic field goes in a helical path around a magnetic field line, and also generates some magnetic field itself, so the solar wind can carry away a trapped portion of magnetic field along with it.

The solar wind is not uniform, and it flows nearly radiallly outward, although it is bent by the Sun's rotation. When it reaches the Earth its speed is between 300 and 900 km/second. 900 km/second is about two million miles per hour.

The solar wind extends throughout our solar system and far beyond. Its region, called the heliosphere, goes out 100 or 200 astronomical units, way past Pluto, to the heliopause. An astronomical unit is essentially the mean distance between the Earth and the Sun. The solar wind has magnetic field as well as ionized particles. The solar wind interacts with Earth's own magnetic field to cause the Van Allen belts and the aurora borealis.

The solar wind has removed mass and energy from the Sun for nearly 5 billion years, so the Sun used to revolve much faster than its present period of 4 weeks per revolution.

Many other stars besides our Sun have stellar wind.

# What Country Will Own Mars and the Moon?

What country will own Mars and the Moon? (Soviet astronauts were the first to orbit the Moon and U.S. astronauts were the first to walk there.) One company selling Moon real estate says they filed a claim with the U.S. government, but when Apollo first landed on the Moon the U.S. stated that they weren't claiming any part of it as U.S. territory.

Space Law is a branch of international law similar to the law of the high seas and air law. Air law applies in the Earth's atmosphere, and space law outside the atmosphere, but the atmosphere doesn't have a distinct border. Some satellites stay in the same place over the Earth (geocentric orbits) so in 1975 Colombia and some other equatorial countries claimed that those satellites are in their territories.

The "UN Committee on Peaceful Uses of Outer Space" (1959) is a central forum for developing legal principles for space. An "Outer Space Treaty"(1967) states that outer space is free for use by all states. Other treaties are: "Agreement on the Rescue of Astronauts, Return of Astronauts, etc'(1969); "Convention on International Liability for Damage Caused by Space Objects" (1972); "Convention on Registration of Objects Launched into Outer Space" (1975); and 'Agreement Governing the Activities of States on the Moon and Other Celestial Bodies" (1984). To some of these, the U.S. will never agree.

D. Granqvist's master's thesis at the Stockholm University of Law tells a lot about the legal aspects of this subject.

References:

For "Personality Profile of the Sun"

1. Zeilik, Michael "Astronomy".. John Wiley & Sons, Inc., New York. 7th ed., 1994
2. Chaisson, Eric and McMillan, Steve "Astronomy Today." . Prentice Hall, New Jersey, 1999.
3. Illingworth, Valerie, Editor.. 1994 Illingworth, Valerie, editor: "Facts on File Dictionary of Astronomy" Market House Books Ltd., 3d ed., 1994.

# Chapter 6
# Stars

## The Star of Bethlehem

Mark (an astronomer): Hi. I'm Mark. I'm a visitor to your church. Will you be speaking tomorrow?

Christa (minister): Welcome, Mark. Yes. My sermon on Sunday is going to be about the three wise men. Mathew 2, verses 1 through 16, tell us how three wise men from the East followed a bright star to Bethlehem. They brought gifts of gold, frankincense and myrrh for the infant Jesus. Only later did it become traditional to refer to these worshipers as kings.

Mark: Were they the shepherds who came to the manger the night of His birth?

Christa: No. The wise men came six months later, to a house. The star that guided them on their 450-mile westward journey from Babylon has been analyzed in many modern astronomy articles.

Mark: I know; I'm an astronomer myself and I've been studying the Star of Bethlehem. The most feasible theories are that it was a supernova (an exploding star), a comet (a rocky snowball orbiting the Sun), a meteor (a particle burning up in Earth's atmosphere), or a conjunction of several planets looking like a single bright star.

Christa: Some Bible scholars explain the beacon as a miraculous spiritual light such as an angel, because they realize it would be difficult to navigate precisely to a house by a star that "went before them, till it came and stood over where the young child was."

Mark: Most astronomers don't get into the question of whether the star was natural or divinely created, Christa. Observers in the Orient recorded a supernova in 5 BC near the stars Alpha and Beta Capricorni. A very bright one, this could have been the Star of Bethlehem. Another report, from Korea in 4 BC, was of a supernova in the constellation Aquilla. As for comets, no significant ones were recorded about that time. Halley's comet was visible in 11 BC, but that was too early to account for this. Meteors don't last long enough for trip guidance.

Christa: It's probably hard to know where the stars and planets were two thousand years ago because the original calendars may have been inaccurate and their formats have been changed since then.

Mark: Yes. The various theories about conjunctions of planets require knowledge of just when the events happened, and the year of Jesus' birth isn't certain.

Christa: One clue is that when He was born, perhaps in 5 BCE or before, or even as late as 2 BCE, King Herod was still in power. Incidentally, someone recorded that there was an eclipse of the Moon shortly before Herod died.

Mark: There was a partial lunar eclipse visible in Jerusalem after dark in 4 BC and a total eclipse at sunset in 1 BCE. Also, in June of 2 BCE, which may have been before Herod died, Venus appeared to be so close to Jupiter that the pair would have looked like a single bright star, because people didn't have binoculars then.

Christa: Planetarium managers often present Christmas programs about the Star of Bethlehem because the public wants to know these theories. Are there any more scientific ideas about what the star was, Mark?

Mark: Yes, there are a lot, some of which seem rather unlikely, but another good possibility is that in 7 BCE a close conjunction of Jupiter and Saturn occurred that was pretty dramatic. Partly because of the Earth's own movements, there was an appearance of retrograde motion of the orbits of Jupiter and Saturn that caused them to pass each other three times in rapid succession, about one degree apart. However, Jupiter and Saturn could have easily been resolved by the human eye as being separate, since one degree is twice the diameter of the Moon, after all.

Christa: It's hard to know which of these theories is correct, and the choice is probably up to the individual. Some skeptics even think that Mathew, who didn't write his gospel until years after the event, may have based this part on unreliable reports. Are there any other biblical phenomena that involve astronomy?

Mark: Oh, yes. They include an enigmatic darkness during the crucifixion, Joshua's long day ("and the Sun stood still"), the age of the universe, and several remarkable observations about the Moon.

Christa: I understand that more than half the people on Earth still worship the Moon.

Mark: How about talking some more, Christa, over a drink?

Christa: Wwwelll, uuhh, why not? Coffee. Starbucks.

# Diameter and Brightness of Stars

Occultation  Astronomers can't directly measure the diameter of any star image with a telescope because they are so far away (except the Sun).

The diameters of some of them can be measured by "lunar occultation,"that is, by measuring the time it takes for the Moon to just completely block the star after it starts blocking the star. It is like an eclipse. Of course they can time an occultation very precisely, and they know how fast the Moon is moving relative to the background stars.

Occultation by the Moon (and by planets) is also used to find the dimensions of radio and X-ray objects, and the positions of objects such as distant radio sources. It can even give information about planetary atmospheres.

There are several more complicated ways of finding star diameters.

Brightness  As to the brightness of stars, here are some definitions.

1. Magnitude Scale is used to measure the brightness of a star. The magnitude scale is logarithmic, and a change of 1 magnitude is a change in lumen brightness of 2.5 times. Unfortunately, magnitudes describe brightness inversely so that smaller positive numbers indicate brighter stars, with zero and negative numbers indicating still greater brightness.

2. Apparent Magnitude is how bright a star appears as seen from the Earth, using the Magnitude Scale.

3. Absolute Magnitude is the true or intrinsic brightness of a star. This gives the brightness that

a star would have as seen from Earth if the star were 32.6 light years away. By pretending that stars are all at that same distance, we can compare the intrinsic brightness of different stars.

4. Luminosity is the total energy radiated by the star per second, often expressed as compared to the Sun.

Example. Take the blue star Rigel (pronounced RYE-jel), which is atOrion's left foot. Rigel's diameter is 58 solar diameters. Its apparent magnitude is + 0.12; its absolute magnitude is -7.1; and its luminosity is 55,000 times much as the power emitted by the Sun.

---

## Bright Ideas About Bright Stars

Awesome! A supernova is an exploding star that is very much brighter than our Sun. Over a period of a few months a supernova can emit as much energy as our Sun would radiate in a billion years.

Chinese astronomers recorded a supernova in 185 BCE, and astronomer Tycho Brahe observed one in1572. Johann Kepler saw one in 1604. In our Milky Way galaxy there is a supernova about once per century on average. When they explode they distribute heavy elements throughout the galaxy.

With modern telescopes hundreds of such explosions in other galaxies can be seen every year, which amateur astronomers are very helpful in discovering. Our Sun is too small to become a supernova, and will probably devolve into a white dwarf.

First bright idea. Scientists noticed that one type of supernova pulsates like a lighthouse and that its frequency of pulsation correlates strongly with its intrinsic brightness. Hence that type of supernova, type 1a, serves as a reliable "standard candle." The type 1a is believed to start as a binary pair of stars that runs out of nuclear fuel and implodes, then explodes.

Second bright idea. Astronomers are able to learn how far away a type 1a supernova is by measuring its frequency of pulsation and its apparent brightness. They have to make corrections for space dust that it passes through and other factors.

Third bright idea. Cosmologists are analyzing hundreds of supernovae to learn about the outward acceleration of the universe. (This plural ...vae is pronounced "supernovee.") Some supernovae are dimmer than expected, so they must be farther away than previously thought. Redshifts disclose how fast the star has been moving away from us. Comparing the redshifts of ones nearby with those far away shows how the rate of expansion of the universe has changed over billions of years.

In 1998 astronomers hypothesized that about five billion years ago the universe began to expand faster. After the Big Bang, which occurred about 14 billion years ago, there may have been cosmic deceleration caused by gravitational attraction, followed then by cosmic acceleration due to a phenomenon described by a "cosmological constant".

Fourth bright idea. The "cosmological constant" has been taken out of the trash bin recently to account for the increase in the rate of expansion shown by supernovae (and by evidence of dark matter and dark energy, etc.). It is a term that Einstein added to his theory of general relativity in 1917 to represent a pushing away of every point in space by surrounding points. It acts against gravitational attraction. He thought he needed such a factor because gravity alone would make the universe shrink, and in 1917 scientists were pretty sure that the size of the universe was constant. Einstein later disavowed his cosmological constant, dubbing it his "greatest blunder."

However, in 1929 Edwin Hubble showed that the universe is actually expanding. Now, long after Einstein's 1955 death, the cosmological constant is back in favor, because the expansion appears to be accelerating. Since 1995 many astronomers have come to believe that the universe contains just enough matter and energy (including dark matter and dark energy) to stop expanding eventually.

Supernovae have not only shed stupendous light through all of space, but have greatly illuminated our understanding of the universe.

---

## Who Owns the Stars?

Had ANOTHER birthday! Received a gift of a star. "Know ye herewith that the International Star Registry doth hereby re-designate Pisces RAOh59m1597sD16o-20'0.21" with the name Charles H. Grace." Pisces happens to be my sign. My best friend paid $83 for it. There are some star charts and a certificate, which I framed. It's the heaviest and longest-lasting birthday present I ever got. But it's a modest body, faint (about 10th magnitude), and light-years away,

What a thrill, to be a part of the universe for the next several billion years. It's a euphoric out-of-this-world gift. Twinkle, twinkle, my little star. I don't wonder what you are.

You are my own namesake, to shine away long after I shuffle off. Even if that's all I ever leave, I'm immortalized.

Maybe it's hot property. I don't actually own the real estate, but I have the naming rights, if only within a private company called "International Star Registry."

You might think the USA, being the strongest nation in the world, would own at least the Moon. Sorry, but no. In1967 the United Nations sponsored an "Outer Space Treaty," ratified by 98 countries, saying that outer space is "the domain of all mankind," so no country owns anything out there except their satellites. That treaty didn't prohibit private individuals from owning space, so a fix-up "Moon Treaty" of 1982 had to be made that, inter alia, prohibits private individual ownership of the Moon and other celestial bodies.

The United States, Russia and China refused to sign the Moon Treaty, for military or political reasons. Countries that didn't sign might dispute the authority of the UN to restrict ownership of space.

Some critics say these treaties discourage expansion and innovation in space. Ownership, or least a license, will probably be required if we expect to encourage entrepreneurial exploitation. The Moon Treaty even says that countries that don't have the resources to exploit the Moon must be given the right to have minerals from the Moon. Private investors are naturally leery of that.

The laws of some countries say an ownership claim alone is not enough–there must be "intent to occupy." And compared with the owning of celestial bodies, ownership of empty parts of outer space is a sticky wicket because it is even harder to occupy and to mark the boundaries. Similar issues were raised and worked out about the exploration and colonization of Antarctica.

Nevertheless, Dennis Hope, a space lawyer who thinks big, claims to own the Sun and the Moon. Mr. Hope and other entrepreneurs have made a lot of money selling extraterrestrial property to hundreds of thousands of people around the world. If the UN treaties prohibiting ownership turn out to be invalid, might Mr. Hope charge us for solar energy? Property ownership entails not only rights but responsibilities, so Hope might find he is liable for skin cancer. Some day it might be OK to own property on the Sun, other stars, the Moon, and planets, but the people presently selling them haven't obtained a clear title yet.

Even though I can't visit it, having my celestial time-share may be the closest I'll get to heaven. Its distance doesn't stop me from rhapsodizing about this eponymous gem. The glamour of New York, Hollywood and Paris are there in a single pinpoint. Its white color is like the magnanimous aural spectrum of an orchestra. It scintillates like the vibrato of a violin, is reliable as a crystal, and as steady as massive Gibraltar. Feet of clay nevermore; now the sky's the limit.

The weather's been pretty clear today.
Thoughtless evanescent clouds,
  Please don't play peek-a-boo tonight.
Don't come between, unwelcome shrouds,
  Nor shade that cherished speck of light.

---

## Why Is a Supernova So Fascinating?

Supernova: An astronomical phenomenon in which most of the material in a star explodes. There is a supernova about every hundred years in a typical galaxy. It is very bright, emitting tremendous amounts of energy in a few months.

Because a "b/pe Ia" supernova is a reliable standard candle, astronomers are able to learn how far away it is by measuring its brightness with a photometer. Dust effects are factored out. They make corrections for the shape of the curve of light versus time, and for color. Astronomers are analyzing supernovae to learn about the outward acceleration of the universe.

Redshifts of Supernovae: The redshift of a supenova discloses how fast it was moving away from us. (The wavelength of radiation coming from a supernova appears greater because of cosmic expansion that occurs while the radiation is in flight.) A Hubble diagram can be drawn for a group of supenovae.

Acceleration: The redshifts of different supernovae, nearby and far away, give astronomers information about the different rates of expansion of the universe over billions of years. They now think that the rate of expansion of the universe began to speed up about five billion years ago. After the Big Bang, there may have been cosmic deceleration caused by gravity, followed by cosmic acceleration caused by the cosmological constant effect.

Cosmological Constant: This is a constant that Einstein added to his theory of general relativity in 1917 to represent a pushing away of every point in space by surrounding points. It acts against gravitational attraction. In1929 Edwin Hubble showed that the universe is expanding, and the cosmological constant. is being brought back.

Critical Density: If there is just enough matter and energy to decrease the rate of expansion of the universe eventually, the universe is said to have at least critical density. According to Einstein's theories, matter and energy are interchangeable, so the newly discovered "dark matter" and "dark energy" contribute to density, along with other matter and energy. Since 1995 astronomers have come to believe that the universe does have critical density.

## Cepheid Pulsating Stars

American astronomer Henrietta Leavitt (born 1868), studied stars that pulsate in brightness. She discovered about 1910 that the period of pulsation of a Cepheid (SEE-fee-id) star is related to the star's average brightness; brighter stars have longer periods of pulsation. The periods are mainly in the range 1 to 50 days.

A Congregational minister's daughter, Henrietta Leavitt attended Oberlin College and graduated from Radcliffe, after which she became deaf from a disease. While working at Harvard College Observatory she measured thousands of variable stars. In 1912 she published results verifying that the brighter pulsating stars seemed to have slower pulses, i.e., longer periods. They were the Cepheid pulsating variable stars. She said "There is a simple relation between the brightness of the variables and their periods." Most of the stars she studied were at about the same distance, so their intrinsic luminosity was attenuated the same amount by their travel, and their brightness as seen at Earth could be compared from star to star.

Luckily, a year after Leavitt's report, Ejnar Hertzsprung determined the distance of several Cepheid stars in the Milky Way without reference to their pulsations, and this information could later provide a calibration by which the distance to Leavitt's Cepheids could be determined.

In 1914 the American astronomer Harlow Shapley got the great idea that Leavitt's relationship between period and luminosity could be used to determine the distance to other pulsating stars. Shapley, born on a farm, dropped out of school to work, but returned to complete a six-year program in only two years, and graduated from high school as class valedictorian.

Shapley measured the period of a Cepheid, and that told him its intrinsic luminosity because of Leavitt's work. He then measured the apparent brightness of that Cepheid with a photometer. By comparing the apparent brightness with the intrinsic luminosity he calculated the distance to the star, based on inverse square attenuation.

In the early 1900s it was not clear that millions of visible stars were actually in other galaxies, as was later to be shown by Edwin Hubble. Their distances could not be measured in the early 1900s, but Cepheids later provided a way, thanks to Leavitt and Shapley. Cepheids were soon detected in other galaxies such as the Andromeda Galaxy, and their distances were measured and found to be enormous. Cepheids showed that other galaxies exist far beyond our Milky Way. Leavitt's discovery changed our model of the universe.

Cepheids are a large group of bright yellow super-giants. The "prototype" of classical Cepheids is a pulsating star called Delta Delphei, which was discovered In 1784.

In 1918 English cosmologist Sir Arthur Eddington (a Quaker pacifist), devised a theory that would explain such pulsations. A star pulsates if it isn't in "hydrostatic equilibrium." Its gravitational force doesn't exactly equal its outward gas-and-radiation pressure, and the outward pressure makes it expand. But because of inertia it doesn't stop when it reaches an equilibrium radius that would exactly balance the gravitational pull.

It overshoots and it slows under the influence of gravity until the outward momentum is zero, after which it starts contracting. Again it overshoots the equilibrium radius and continues to shrink until the gas pressure within the star finally reverses it. The star is pulsating in size.

As the size changes, the intrinsic luminosity also changes, according to Eddington's theories, so the observed brightness pulsates. Eddington also showed that the period of pulsation is related to the star's average radius and mass. The luminosity of the star also depends upon the average radius and mass of the star. Thus the period of the pulsation is related to the star's average intrinsic luminosity.

Henrietta Leavitt's work was the basis for Edwin Hubble's proof that there are other galaxies. She greatly changed modern astronomy, although she didn't receive much credit during her 53-year lifetime.

## References:

For "The Star of Bethlehem

1. Bouw, Gerardus "The Biblical Astronomer," vol. 8, no.86, p. 12, 1998.
2. " What Was the Star of Bethlehem?" Mysteries of the Universe. www.msnbc.msn.com.
3. Unruh, J. Timothy "The Star of Bethlehem." Biblical Astronomer, no. 2. .
4. Molnar, Michael R., "Revealing the Star of Bethlehem." www.eclipse.net/molnar.
5. Avalos, Hector, "The Bible and Astronomy." www.secularhumanism.org

For "Bright Ideas About Bright Stars"

6. NASA.gov. Imagine the Universe. Supernovae.
7. NASA.gov. Heasarc: Education & Public Information. Supernova.
8. Wikipedia.org. Supernova.

For "Who Owns the Stars?"

9. Daintith, John and Gould, William "Cepheids." The Facts On File Dictionary of Astronomy, , Infobase Publishing, New York.
10. Wikipedia, the free encyclopedia "Harlow Shapley."
11. Wikipedia, the free encyclopedia "Arthur Eddington."

# Chapter 7
# Planets

## Humans on Mars in 2048

"Vickie, your Mom and I only want you to be happy. We're worried about your serious dating with that boy Raymond who's visiting from Mars."

"What's the problem, Dad-- I'm 17 years old. It's 2048, not 2010; I hope you aren't living in the past. Michael comes from a good family. His parents were selected to be colonists back in 2030 by NASA because of their good genetic qualities."

"Yes, but Mars has a very different subculture and life-style, honey. It's almost as different as in that other human branch, the Moon people, and Mars is farther away."

"Mom and Dad, Mars inspires me! Pioneer women helped to develop North America."

Why Colonize?

If we colonize Mars, and an asteroid wipes out all life on Earth, colonists can come back and repopulate the Earth. Space colonization may enable the human species to survive for another million years. We can't build a Mars colony right now, but probably will within the next few decades.

The probability of extinction of almost all life on Earth may seem remote, but it happened when the dinosaurs died. We are studying techniques to deflect incoming asteroids, which may or may not be successful.

What would life be like on Mars?

Mars is more like the Earth than the Moon is. A day on Mars is almost the same length as a day on Earth, i.e. 24 hours 31 minutes of Earth time.

The Martian year is 1.88 Earth years, Mars being 142 million miles from the Sun. At the right time in our orbit, Mars is only 40 million miles from Earth. A spacecraft takes about nine months to get there, which is a long time in an emergency, and costs about $5 billion. The travel time will no doubt be reduced greatly with further engine research.

Mars' axis of rotation is inclined 25.2 degrees from the plane of its orbit around the Sun. (Earth's tilt is 23.4 degrees.)

The atmosphere on Mars reduces its susceptible to catastrophes from micro-meteorites and Sun damage. Even though Mars' atmosphere is thinner than Earth's, the radiation at the surface is about the same as on Earth because Mars is farther from the Sun. Mars' orbit is much more eccentric than Earth's, so the effects of sunlight vary greatly throughout its year.

Mars is half the radius of Earth and one-tenth the mass, so gravity there is 38% of Earth's gravity. That may be enough, barely, to avoid gravity-deprivation illnesses such as poor fetal development and bone loss.

The mean surface temperature is a frigid minus 38 degrees F., with a low of minus 252 degrees F. Mars has a lot of minerals and other natural resources needed to construct a civilization, probably including water ice. There may be caves to live in, and it may even be possible to grow some crops.

What would life be like on the Moon?

The Moon is only 240,000 miles away, so a spacecraft can get there in a couple of days. Round-trip flights might cost about $240 million.

It is about one fourth the diameter of Earth, and has gravity of only one-sixth that of Earth. Although it doesn't appear to have any fertile land and is exposed to solar flares, and has no bees for pollination, some limited crop growth may be feasible.

Because the Moon always presents the same face to the Earth, a Moon "day" is the same as its time for orbiting the Earth, i.e., 28 days. That means 14 Earth-days of hot bright sunlight (225 degrees F) followed by 14 days of bitter cold darkness (minus 243 degrees F.). There is no significant atmosphere so there isn't much protection from the Sun or meteors.

Fortunately, the axis of rotation of the Moon is almost perpendicular to the ecliptic. A colony located near the north or south Pole of the Moon would have much less extreme changes of weather than at other latitudes, similar to the seasonal long days and long nights of northern Alaska. The average temperature at a Moon pole is minus 58 degrees F. We can live with that.

A Moon colony could mitigate the extreme temperatures if it were at least 12 feet underground, where the average temperature is 74 degrees F. A surface-mounted colony in modules covered over by Moon-dust might accomplish the same thing.

The Moon would be a marvelous site for astronomical observatories, partly because it rotates so slowly and is almost free of man-made interference.

Tourism and research could be big financial benefits to a Moon colony.

Which should be colonized first–Mars or the Moon?

This is a hot controversy that will probably be resolved by settling on the Moon first and using that as a launching station for colonizing Mars. The USA landed humans on the Moon temporarily years ago, and a resident colony would be the next step. The European Space Agency is shooting for 2025.

A Moon colony might be more salable politically than a Martian one because it seems more feasible to the public. Any colony would have to have a politically correct diversity of residents.

Opponents point out that for what it will cost to colonize Mars or the Moon we could pay for a lot of food stamps for hungry people here on Earth. On the other hand, some scientists argue that the world should be spending ten times as much on space exploration as the $50 billion a year that it has been, which would still be only a quarter of one percent of the world's gross domestic product.

"Well, Vickie, please think about this carefully. Raymond may want you to move to Mars. You don't know how hard it was for me just to change from my family's religion when I wanted your Mom to marry me. Try to picture how hard it would be to adapt to life on Mars after growing up in Ohio."

---

## Why Is Mars So Close Sometimes?

In 2003 two things were happening at the same time: (a) Earth and Mars were in opposition, i.e., in the same direction from the Sun, and (2) Mars was near its perihelion. There were also several less important contributing factors, one of which was that the Earth was not far from its aphelion, which occurred in July 2003.

All of the planets have elliptical orbits; so said Kepler almost 400 years ago. Earth's distance from the Sun varies about 2%, and Mars' distance from the Sun varies about 9%. As a result, the distance from Earth to Mars, at opposition, can be as little as 34.6 million and as much as 62.8 million miles.

Opposition of Earth and Mars occurs every 26 months, but it is rare that Mars is at its perihelion at the same time as the opposition with Earth. This happens rhythmically every 15 to 17 years.

Mars was about 34.6 million miles away in September of 2003. It hadn't been that close since Neanderthal times, 60,000 years ago. There will be another close pass in 2018, but Mars won't again be as close as it was in 2003, until 2287. That's true even though Mars' orbit is slowly becoming more eccentric because of the pull of Jupiter and other bodies.

When will the aphelion of Earth line up in opposition with the perihelion of Mars? The perihelion of Mars advances about 5.5 arc seconds per century, and the aphelion of Earth advances about 10.4 arc seconds per century; it will be thousands of years before Earth's aphelion catches up with Mars' perihelion.

At the beginning of September, 2003, Mars rose at nightfall in the southeast and reached its highest point at about midnight. That's when it was easiest to see because there was the least atmosphere in the optical path. Mars is in the constellation Aquarius now.

With a telescope having a mere 75 magnification, Mars looks about as large as the full Moon does to the naked eye (1/2°). It is red or orange, but sometimes kind of yellow, and very bright (magnitude reaching minus 2.9).

Incidentally, Earth and Mars are alike in many ways. A day on Mars is only a half-hour longer than an Earth day. The axis of rotation of each is inclined about the same amount from its orbit, so their seasonal variations are similar (but Mars is much colder than Earth). They are both terrestrial planets, i.e., rocky rather than gaseous. As seen from the celestial zenith, Earth and Mars both revolve counterclockwise in their slightly tilted orbits.

But in other ways Mars is very different. It is only 4200 miles in diameter. Mars is one and a half times as far from the Sun as the Earth is, so Mars' sidereal period is about 1.9 Earth years.

---

## Numbers in the Solar System

- 6 billion years ago. Our solar system began to form from a cloud of interstellar material.
- 880,000 miles. Diameter of the Sun--more than 100 times Earth's diameter of 8,100 miles. The Sun contains 99.9% of the mass of the solar system.
- 8 The number of full planets in our solar system now that Pluto has been downgraded to "dwarf planet." In business, someone who has been demoted is now said to have been plutonized.
- 3/8 Diameter of Mercury compared with Earth. The smallest planet, Mercury is the closest to the Sun and has the most eccentric orbit.
- 11 times. Diameter of Jupiter compared with Earth.
- 318 times. Mass of Jupiter compared with Earth. The most massive planet and the fifth one from the Sun, Jupiter has 16 Moons.

- 23 hrs 56 min. Earth's period of rotation on its axis. Add 4 minutes to complete the rotation
- 165 Years for Neptune to go once around the Sun (Earth years).
- 30 AU. Distance of Neptune, the farthest planet, from the Sun. An astronomical unit (AU) is the distance of Earth from the Sun, i.e., about 93 million miles.
- 8.3 minutes. Time for sunlight to travel to Earth.
- 4 Number of planets in our solar system that probably have water, that life-giving elixir. They are Venus, Earth, Mars and Jupiter.
- 5 km/sec. Escape speed from Mars. Earth's escape speed is 11.2 km/sec. It doesn't depend on the mass of the object that's escaping.
- 6 times. Reflective power of Earth compared with the Moon. The Moon's albedo is only 0.07, but Earth's is 0.40 because Earth has white clouds.

---

## Planet Hide and Seek

I thought I saw a putdy cat. Is a planet lurking there? Acting as detectives, astronomers meticulously examine every obscure clue. But unlike with DNA, they're never very sure.

A planet is a celestial body that generates no light of its own, but is illuminated by a star around which it revolves. The reflection is extremely faint compared with its parent star, whose glare makes the furtive planet hard to see.

One way of detecting planets is the "Radial Velocity Method." A star and planet revolve around their mutual center of mass, called the barycenter. The barycenter is offset from the center of the star (although it often falls within the diameter of the star); and the star moves in its own small orbit around the barycenter. Doppler measurements to the rescue! As the star orbits its barycenter, a component of its orbital velocity may be parallel to the line of sight to the Earth. Hence the spectral lines of the star have Doppler frequency variations. Those clues give away the diameter of the star's orbit!

However, only the minimum possible value of the mass of the planet is disclosed this way, because the planet's orbit may be almost perpendicular to the line of sight to the Earth, making the star's Doppler effect smaller by a cosine factor.

Another search technique is the "Transit Method." If a planet passes in front of its host star it blocks some of the light, as seen from Earth. Photometers detect the periodic light reduction, with the percentage of blockage depending upon the areas of the planet and the star. While most other methods show only the mass of the planet, this method also indicates its diameter.

Other schemes in the planet quest include:

Direct Imaging of the planet, which is useful only for a planet that is very large, young (hence hot, with strong infrared radiation), and widely separated from its parent star. It is most feasible if done from a spacecraft. Sometimes the bullying image of the host star can be masked out.

Astrometry, in which the angular position of a star (not its Doppler) is measured with such precision that its orbit around the barycenter can be plotted. Even the best Earth-mounted telescopes aren't discerning enough to use this method, but in 2002 the Hubble Space Telescope confirmed that one previously suspected planet does indeed exist.

Pulsar Timing. A pulsar is a remnant of a supernova, and they are rare. It sends out radio pulses with extremely regular timing. If a pulsar has a planet, when the pulsar moves in its small orbit around the barycenter, the timing of its pulse varies infinitesimally. In1992 Alexsander Wolszczan and Dale Frail, using this method, discovered the first confirmed planet outside our solar system. That particular planet can't sustain any life because its pulsar constantly sprays it with high-energy radiation.

Gravitational Microlensing. This clever idea depends on the curvature of space near a mass. When the images of two stars are closely aligned, even if the stars are at greatly different distances, the "gravitational field" of the foreground star curves the neighboring space enough to act as a lens that magnifies the background star. If the foreground star happens to have a planet, that planet's "gravitational field" also affects the lensing, so the existence of the planet can be detected. Polish astronomers found several possible planets this way in 2002, but their discoveries can't be confirmed because the alignment of those stars will never happen again.

Some of the planet-detection ideas being worked on for the future are called by the strange names "orbital phase reflected-light variations," "eclipsing-binary minima timing," and "polarimetry." Esoteric, yes, but very promising.

## Finding Extrasolar Planets

Two fast-moving dramas:

1. Mankind is uncovering nature's secrets about planets outside our solar system.
2. Competing scientists are using various ways to find the planets.

CNN news said on September 17, 2002, "Astronomers scanning southern skies in search of distant solar systems have discovered a Jupiter-like planet 100 light years away circling a star similar to our own Sun....To find evidence of planets, scientists use a high-precision technique that watches for stars that appear to wobble because of the effect of an associated planet's gravitational pull. ... The wobble can be detected by the Doppler shift it causes in the star's light, indicating the presence of a planet. This is then fine-tuned to assess its distance and mass." The newly found planet is in the southern constellation Grus (the Crane).

Astronomers have found about a hundred such planets outside our solar system. One type of planet orbits very close to its star, and another type orbits much farther out. This new one is of the farther-out type, in which the orbits are more circular. It is about three times as far away from its star as the Earth is from the Sun.

Researchers are making a list of analogues to our solar system for future space trips. Later they'll look at them for signs of life, such as carbon, carbon dioxide, ozone, etc.

Three of the ways of searching for extrasolar planets are:

1. Radial Velocity Searches. This popular method, which a team used to find the planet in Grus as described above, detects very small radial motions of the star by using large telescopes and precise spectrographs to observe redshifts and blue shifts. They used the Anglo-Australian telescope in New South Wales.
2. Photometric Searches. The "Arizona Search for Planets" team likes this method. They measure the reduction in total light from a star as its planet passes in front of it. The planet blocks a small part of the

light from the star, so a precision photometer sees a downward square wave of very small amplitude while the planet is traversing the star.

3. Astrometric Searches. A star that is being circled by a planet can trace out a very small circle in the sky because of the gravitational pull of the planet. The circle might be seen by tracking the angular position of the star. This method isn't very good because the circle is so small that it is very hard to detect.

---

## Terrestrial Planet Finder

Terrestrial Planet Finder (TPF) is a NASA project to build a telescope system for finding planets outside our solar system. A "terrestrial planet" is one that is small, rocky, and has a shallow atmosphere, like Mercury, Venus, Earth and Mars, as contrasted with large gaseous planets. Astronomers hope the project will find planets that are habitable, with atmospheres and chemical characteristics conducive to life.

The main problem is that a terrestrial planet shows only weak reflected light and is always located near a bright host star. It's like trying to see a firefly near a headlight on a distant car. Resolution and sensitivity are also difficult. Two kinds of apparatus are being worked on.

Visible Liqht Coronaqraph

This will be a single telescope with a mirror three or four times larger than the Hubble and more precise in its construction. It will tremendously attenuate the star's component of light so that planets can be seen. The image of the bright star will be blocked out by an occulting disc located at the prime focus of a refracting lens. A refracting lens doesn't scatter light as much as a mirror. Coronagraphs were invented in 1930 by a French astronomer to look at the Sun's corona when there isn't a total solar eclipse.

Infrared Astronomical Interferometer

This will be an array of spaced-apart telescopes, each perhaps 10 feet in diameter, which may extend about 100 feet. They will receive infrared signals and combine them in such a way as to achieve the resolving power of a large single telescope of 100-foot diameter. The array will be orbiting the Earth extremely far out in space, perhaps 90 million miles away, and could be either a rigid structure or a group of spacecraft flying in formation. This interferometer approach will use a technique that nulls out the starlight by a factor of one million so the infrared emission from its neighboring planet can be seen. Nulling techniques are daunting and extremely complicated. Locally-generated light signals, probably from lasers, might cancel the light from the star and other extraneous light sources. To detect a planet the light processing has to be done at cryogenic temperatures, probably below minus 230 degrees Centigrade.

The star Rigel Kentaurus has the highest priority for being a TPF target. It is in the Centaurus constellation of our southern hemisphere, which is only four light-years away and partly in the Milky Way. Another promising target star is Tau Ceti (seen near Orion), which is twelve light years distant. Other candidate targets are as remote as seventy light years.

In February 2006, NASA's budget for 2007 postponed the TPF project indefinitely, but five months later some funding was approved, and the project should go forward. The systems are targeted to start operating between 2014 and 2020.

The European Space Agency is working on a similar project, which they call Darwin.

## Saturn and Its Spectacular Rings

It has those gorgeous rings, it's big, it's far, and it's light-weight. Saturn takes 29 years to go around the Sun, so about every 15 years the rings are edge-on and all but disappear (they are less than 2 km thick). That happened in 1994-95.

One-quarter of the orbit later the rings are the most visible from Earth. And of course they are most easily seen when the Earth and Saturn are on the same side of the Sun (in opposition).

Its seven rings consist of thousands of individual particles, each of which is actually a satellite of Saturn. The particles are up to several centimeters in size, but on ring B they are even a few meters.

Where is it? Saturn is almost 10 times as far from the Sun as the Earth is. It is the sixth planet from the Sun, between Jupiter and Uranus.

How big? Saturn is the second largest planet in our solar system (after Jupiter), with a diameter of 121,000 km. That's almost ten times Earth's diameter, which is only 12,800 km (8,000 miles).

How massive? Saturn's mass is almost as great as that of the Earth. It has the lowest density of all of our planets, being only 70% as dense as water. Earth's density (at 5.5) is eight times that great.

What is it made of? Hydrogen seems to form most of its mass, but like Earth, Saturn probably has an iron-rich rocky core.

Rotation on its axis? Its axis is tilted 27°, which is comparable to the 23.4° of the Earth. It rotates on its axis once every 10 hours, 39 minutes, that is, more than twice as frequently as the Earth.

What satellites does it have? Although Earth has only one natural satellite (the Moon is 2000 miles in diameter), Saturn has 18. They include Titan (3200 miles in diameter) and Rhea. Eight of its satellites were discovered since 1978.

How bright is it? The average apparent magnitude at opposition is 0.7. Saturn's geometrical albedo (apparent reflectivity) is 0.47; the Earth's is 0.37, for which our clouds get most of the credit.

Inspection visits? We have sent a couple of probes out to Saturn–Pioneer II in 1979 and the Voyagers in 1980 and 1981. They sent back a great amount of information about Saturn's meteorology and atmosphere, which are similar to those of Jupiter.

## Shocking Ratios

30 to 1. Distance from the Sun to Neptune (9,000 million km, which is 30 Astronomical Units) compared with the distance from the Sun to Earth (300 million km or 1 AU).

5 to 1. Albedo of the Earth (0.35) compared with albedo of the Moon (0.07). White clouds reflect about a third of the light that strikes the Earth.

2.5 to 1. Reduction in apparent brightness from a star of magnitude 1 to a star of magnitude 2. (A first-magnitude star is brighter than a second-magnitude star.) To a human eye, apparent brightness is logarithmic. The ratio of apparent brightness between a star of magnitude 1 and a star of magnitude 6 is 2.5 raised to the fifth power, hence about 100 to 1.

900,000 to 1. Speed of light (300,000 km/second) compared with the speed of sound (0.33 km/second) in air. Light is almost a million times as fast.

43 to 1. Resolution of the naked eye (60 arcsec) compared with resolution of a 10-cm telescope (1.4 arcsec), for visible light. Diffraction in telescopes causes fuzziness that makes two close objects merge into one. Diffraction is wavelength-dependent, resolution of shorter wavelengths being better.

727 to 1. The U.S. Gross Domestic Product ($12 trillion) compared with NASA's budget ($16.5 billion) for 2006. NASA's budget was one-seventh of 1% of the GDP.

Infinity to 1. (or infinity to any number). Number of Dobsonian telescopes sold in the U.S. in 2010, compared with the number sold before 1960. American monk John Dobson invented them about 1960. A Dobsonian telescope is a Newtonian, so it has a concave collecting mirror at the rear of the tube and an eyepiece on the side. Its alt-azimuth mounting structure is like a cannon that can swivel up and down, left and right. Popular for large-aperture, short-focus reflectors, it is inexpensive, simple, stable and easy to use.

1 to 1. Ratio of the diameter of Uranus (51,200 km) to the diameter of Neptune (50,600 km). The accent is on the first syllable of the word Uranus.

18 to 1. The elapsed time (18 centuries) between the work of Ptolemy (125 AD) and that of Einstein (early 1900's) compared with the time between Einstein and today.

References:
For "Humans on Mars in 2048"
1. Shiga, David. "Stephen Hawking Calls for Moon and Mars Colonies". New Scientist, 2008.
2. "Lunar Outpost" NASA's Plan to Construct an Outpost between 2019 and 2024. NASA.
3. Zubrin, Robert "The Case for Colonizing Mars" The National Space Society. 1996.
4. Dinkin, Sam. "Colonize the Moon Before Mars" The Space Review. 2004.
5. "Colonization of the Moon" Wikipedia.org/wiki. 2009.
6. "Colonization of Mars" Wikipedia.org/wiki. 2009.

For "Planet Hide and Seek"
7. http://en.wikipedia.org

For "Terrestrial Planet Finder"
8. Technical Report: http://planetquest jpl.nasa,gov/tpf
9. Illingworth, Valerie, editor.."Facts on File Dictionary of Astronomy" Market House Books Ltd., 3d ed., 1994.
10. http://en.wikipedia.org

# Chapter 8
# Recent Astronomers

## Edwin the Logical–Hubble's Thoughts

Five hundred years ago in France, some of the Dukes of Burgundy, who were generous art patrons, gave themselves names like Phillip the Good and John the Brave. If we followed that practice today, we could call Hubble "Edwin the Logical." He made several important discoveries, all of which arose out of simple syllogisms.

One of his discoveries was that not all of the stars in the universe reside in the Milky Way. In fact, there are billions of other galaxies, each having billions of stars of its own. He deduced this by looking at a type of stars called Cephied variables. These stars pulsate regularly, with periods of 1 to 50 days. Previously, by observing many Cephieds that were about the same distance away, astronomer Henrietta Leavitt had discovered that their frequency of pulsation correlates with their brightness. Brighter Cephieds pusate faster.

Hubble was studying Cephied variables with the 100-inch telescope at the Mount Wilson Observatory in California when he noticed that one rather dim star was pulsating very fast. That's backwards; only very bright stars are supposed to pulsate that fast. Hubble said—that star might appear dim because it is much farther away than we think. He measured its apparent brightness and decided that it is too far away to be in our galaxy. There must be another galaxy out there. That was logical, and logic was Hubble's long suit.

Until then, astronomers had thought that all of the stars were in our galaxy. The Milky Way was thought to be the entire universe. Hubble said, no, there are many other galaxies farther out there. Evidence of that was one his greatest accomplishments.

Edwin Hubble was born in Missouri in 1889, and read Jules Verne and other science fiction writers as a child. They moved to Chicago, where he was remarkable for his athletic ability in school. He broke the Illinois State high jump record and was an outstanding basketball player at the University of Chicago.

Edwin won a Rhodes scholarship to Oxford, where he earned a law degree. Back in the US, he taught high school for a year, and then decided he wanted to be an astronomer. He obtained a PhD in astronomy at the University of Chicago.

Another of Hubble's great accomplishments was to devise a classification system for the many galaxies he observed. He sorted them by their star content, distance from the Earth, brightness, and shape. The main shape categories are elliptical, spiral, and barred spiral; he also defined subdivisions of each. That classification system is still used today, more than two generations later.

While classifying the galaxies, Hubble noticed that the Doppler shift of their light depends on their distance. The Doppler shift, or redshift, is greater for galaxies that are farther away, which means that they are moving faster. He quantified that phenomenon by drawing a scatter diagram showing that a galaxy's radial speed is proportional to its distance from the Earth.

From that diagram, he proposed Hubble's constant, which is the ratio of the star's speed to its distance from the Earth. His numbers were off by a factor of ten, but the concept was right. Hubble concluded that the universe is expanding. Hubble didn't do any of the complicated math that Kepler, Newton or Einstein had done; he just put two and two together.

If the universe is expanding, it must have started at some time. George Gamow and other US physicists went backward in time and came up with the Big Bang theory. The reciprocal, 14 billion years, of Hubble's constant, is about how long ago the Big Bang occurred. Incidentally, it now appears that the rate of expansion is increasing due to dark matter and energy.

Many things are named after Hubble, including a lunar crater, a minor planet, Hubble's law, Hubble's constant, and a space telescope. In 2002, astronomers using the Hubble Space Telescope discovered a new world beyond Pluto.

Edwin Hubble died in 1953. He became one of history's greatest astronomers by applying simple logic to astronomical observations. Hubble richly deserves to be a Duke in the realm of astronomy, in which case he could fairly be dubbed "Edwin the Logical."

---

## Fred Whipple, Astronomer Extraordinaire

What powers a hybrid car? What powers a priest or a political leader? What powered Fred Whipple? He received many honors in astronomy, including awards from Presidents Truman and Kennedy. He must have had some powerful abilities such as intuition, analytical horsepower, and initiative.

It's hard to be sure what his personal strengths were, but we can speculate.

Intuition The "sixth sense" of intuition provides immediate understanding without analysis. Whipple did play his hunches, and he had a magnificent scientific background on which to base his intuition.

- He had right-on insight as to what might be inside of comets. Fifty years ago he proposed a theory that comets have cores that he dubbed "icy conglomerates," which the press nicknamed "dirty snowballs." Other scientists vehemently disagreed with his theory, but it was vindicated in 1986 when the European Space Agency spacecraft Giotto penetrated Halley's comet and photographed its interior.
- He theorized that most meteors in our solar system originate from the break-up of comets, and that their tails are directed away from the Sun because of the solar wind (a flow of mainly protons and electrons from the Sun's corona).
- He recognized that small changes in the orbits of comets result from jet streams that outgas from their core because of solar heating.
- In WW II Whipple developed countermeasures to confuse enemy radar, including the jettisoning of strips of aluminum from our aircraft.
- He invented the Whipple Shield, which is still used today. It is a thin metal cover that vaporizes small particles when they strike a spacecraft,

Analytical Ability Fred had natural aptitude for mathematics and physics. Perhaps he was able to focus exclusively on one thing at a time. Because of his obvious talents, his parents moved from Iowa to California for him to go to college. He majored in mathematics as an undergraduate, and after taking an introductory course in astronomy he decided that's what floats his boat. He received a doctorate in astronomy in 1931.

- Whipple was an expert on meteors, meteorites, comets and planetary structure.
- His analytical articles about "dirty snowballs" were said to be scientifically brilliant and were the most cited papers in the past 50 years.
- As Director of the Smithsonian Astrophysical Observatory he helped to develop a network of Baker-Nunn cameras (similar to Schmidt telescopes) that was so precise in tracking satellites optically that the exact shape of the Earth could be ascertained from satellite orbit data.
- He pioneered large low-cost multiple-mirror telescopes that are used in an observatory in Arizona named after him.

Initiative Whipple must have been pro-active. As an Iowa farm boy until age fifteen, he probably developed a strong sense of responsibility and of cause and effect. Apparently he was in good health, as his long life would indicate, so maybe he didn't suffer until he was in his 90's from the aches and pains that rob a person of initiative. He died in 2004 at age 97.

- He organized a world-wide "Moonwatch" network of amateur astronomers to track artificial Earth satellites even before there were any, and his group determined the orbit of the satellite Sputnik when it was launched in 1957.
- He discovered six comets, several of which are named after him.

- Whipple wrote several books on the solar system, including a popular one called "Earth, Moon and Planets."
- At age 93 he helped to plan NASA's Contour mission, which unfortunately was aborted for lack of funding.
- He worked at Harvard from 1931 to 1977, and was a professor of astronomy there from 1950 to 1977, becoming chairman of the department.

Whipple didn't devote all his zeal to outer space. He married his first wife, Dorothy Woods, in 1928, and they had one son, but divorced in 1935. He and his second wife Babette Samelson were married nine years later; they had two daughters.

What made Fred Whipple tick? What personal qualities made him so prodigious? Who knows. But humankind is their grateful beneficiary!

---

## Amazing Stephen Hawking

"With Roger Penrose, an English physicist and mathematician, he showed that Einstein's general theory of relativity implied space and time would have a beginning in the Big Bang and an end in black holes."

That quotation is from Stephen Hawking's website, www.hawking.org. Hawking is the English cosmologist who has ALS (Lou Gehrig's disease). Although his mind is still OK, his body is almost completely paralyzed. And we think we have problems! He started showing ALS symptoms when he was a young man. Shortly after that he married Jane, a graduate student, and they have three children. He is now 60 years old and has a grandchild.

Professor Hawking occupies the mathematics and theoretical physics chair at Cambridge that Isaac Newton held in 1663. His accomplishments in cosmology include a great many books, articles, and lectures, including:

- A Brief History of Time (book)
- Black Holes and Baby Universes (book)
- The Universe in a Nutshell (book)
- Inflation of the Universe
- Gravitational Entropy
- Quantum Cosmology,
- M-Theory and the Anthropic Principle

At one time he almost died of pneumonia, and in 1991 he was struck by a car while crossing a street at night in his wheelchair.

His ALS is slowly getting worse. He had to have his larynx removed, and he communicates with an American-made artificial speech system. It has a monitor that shows an alphabetical list of words; he can find and select a word by means of switches. After he has sequentially stored as many words as he wants, e.g., a complete sentence, he starts a speech synthesizer that pronounces the sentence aloud. The synthesizer is easy to understand, and in fact sounds pretty good.

Stephen now travels all over the world to lecture and attend scientific meetings. In September 2002 he received a new computer system from Intel and an elegant new wheelchair from Pride Mobility.

Before he got sick he didn't care much about doing anything. Would he have accomplished as much if he weren't disabled? Did it motivate him to concentrate intensely on cosmology? Did his wife Jane inspire him to apply himself?

Do his admirers exaggerate his scientific accomplishments because he is disabled? I don't know.

We do know that Stephen had ample opportunity just to vegetate and be cared for as a disabled man. What he did instead shows a marvelous spunky drive to help himself. Hawking is an inspiring role model for other disabled people.

---

## More about Stephen Hawking

I used to assume that Stephen Hawking's fame was because he kept working despite being severely disabled. I now find that he is regarded by some as a brilliant theoretical physicist.

Besides his esoteric scientific articles he also writes books that simplify astronomy and cosmology. His first popular book "A Brief History of Time,"1988, was a multimillion bestseller for many years. In 2001 he published "The Universe in a Nutshell" to include scientific developments that occurred after the first book came out. In 2008 Stephen Hawking and Leonard Mlodinow published a paperback called "A Briefer History of Time."

Hawking is severely disabled by amyotrophic lateral sclerosis, or ALS, also known in the US as Lou Gehrig's disease. He was born in England in 1942, where his father Dr. Frank Hawking was a research biologist and his mother a political activist. He enjoyed riding horses when he was young, and he coxwained a rowing team at Oxford where he studied astronomy. He later went to Cambridge to study theoretical astronomy and cosmology, where he received a PhD.

Symptoms of ALS first appeared while he was still a student at Cambridge. He lost his balance and fell, hitting his head.

Worried, he took a Mensa test to verify that his mental abilities were still good. When Hawking was 21, shortly before his first marriage, his disease was diagnosed as ALS, and doctors predicted that he would survive for only two or three years. Language student Jane Wilde married him anyway.

In collaboration with Roger Penrose and others Hawking has proposed several theories about singularities in the framework of Einstein's general relativity. He predicted that black holes should thermally create and emit subatomic particles, sometime called Hawking radiation. With others, he wrote a mathematical proof that any black hole can be completely described by its mass, angular momentum and electric charge. He tackled a wide variety of problems in cosmology.

He gradually lost the use of his arms, legs, and voice, and is now almost completely paralyzed, but still working. He talks with a computer-generated voice. Stephen credits wife Jane Wilde for reviving his will to live. They had three children, but were later divorced. His second wife was his nurse, and they are also now divorced. He has since reconciled with his first family. Stephen doesn't write much about religion; he appears to be an atheist or a pantheist.

Hawking has become famous the world around and received many honors for working under such extreme adversity. Moreover, he seems to know the sweet uses of publicity, often being a subject in books, newspapers, magazines, TV, radio and on the lecture circuit.

Amazing Stephen Hawking is not only a significant scientist, but a role model. He may inspire millions of disabled people to soldier on.

# Competition Among Astronomers

A satellite telescope called the Kepler, to be launched by NASA in March or April of 2009, will revolve around the Sun instead of around Earth. Its solar orbit will make it easier to study the light from distant stars for long periods. The 55 inch mirror of the Kepler can monitor the strength of the light photometrically, of many stars at the same time. It will analyze more than 100,000 stars in its lifetime, which is expected to be 3.5 years.

The plan is to detect small reductions of light intensity that occur when planets cross in front of their host stars. Unfortunately, the orbits of most planets don't cross in front. Of those that do, it may be necessary to watch a star for years before seeing the minuscule light reduction caused by the transit, and scientists want to see it at least three times to be sure it isn't caused by something else. The dip in intensity is typically 100 parts per million, which is 1/100$^{th}$ of 1%.

NASA scientists hope to find planets outside our solar system that can sustain life, and they would love to be the first to find a lot of them. James Fanson and Michael Bicay are among the leaders of the NASA Kepler project.

The Kepler telescope is the latest move in a healthy competition among astrophysicists. Football teams compete, as do boxers, business men, churches for members, men for trophy wives, lions for a harem, and cyclists. Scientists mapping the DNA genome devised ingenious ways to do it faster because they were driven by competition.

Anthropologist Margaret Mead found that some primitive societies value cooperation more highly than competition, and that some even consider competition undesirable. She believed that competitiveness is a culturally created aspect of human behavior. It is pervasive in the United States.

Lawyers compete aggressively. One takes the plaintiff's side, another takes the defense, and both try very hard to be persuasive (but they are severely punished if they lie to a court about the evidence). Courtroom competition is fundamental to the Anglo-American system of jurisprudence. Most legal systems of the world don't stress competition, but we in the USA think it is the best way to get at the truth.

Is competition beneficial or is it a vestigial curse of nature? The struggle for survival may have nurtured competition in antiquity. But even Darwin's theory of evolution does not suggest that competition is the most successful strategy for survival of the fittest.

In 1995 Michel Mayor at the Geneva Observatory found the first planet outside our solar system (exoplanet). Mayor's discovery immediately inspired the French and European space agencies to send a French satellite, called Corot, into orbit to do the same thing from space. Corot is just orbiting the Earth, so it can't continuously keep an eye on a particular star nearly as long as Kepler can. A French woman, Annie Baglin, is the chief scientist of the Corot project.

A few years ago a team of European astronomers using a telescope in the Chilean Andes discovered a planet in the constellation Libra that is remarkably like our Earth. It orbits around a dim red star called Gliese 581 only 20.5 light years away, which is "only" 120 trillion miles. It probably has liquid water and a life-sustaining temperature range. At 1.5 times the diameter of Earth, it has enough mass to retain a substantial atmosphere, which protects it.

Our Milky Way galaxy alone has a hundred billion stars and many of them (perhaps most) have planets. A method of confirming photometric evidence that one of them has an exoplanet is to Doppler monitor the host star, which wobbles because the mass of the encircling planet attracts the host star.

More than 300 exoplanets have been found so far, almost all of them uninhabitable because they are too hot or cold or consist of only gas instead of rock. A climate in which life can exist is sometimes called the "survivability zone," nicknamed the "Goldilocks zone." In our solar system, only Earth fulfills all the survivability criteria.

The inhabitants of exoplanets could look like anything at all. Adherents of various religions are confident that their Gods have jurisdiction over the exoplanets even if the inhabitants are not anthropomorphic.

COMPETITION appears to motivate people very powerfully. Competition seems to be a great way to conduct not only football and business, but astrophysics as well.

---

## What Kind of Person Likes Astronomy?

The three Myers-Briggs personality types that are happiest in astronomy are:

1. The Pragmatist (called type INTJ) who are only 1% of the general population,
2. The Leader (type ENTJ) who are 5% of the population, and
3. The Planner (type INTP), who are 1% of the population, (per Behrens reference below).

Personality types are based on original work of Carl Jung. There are sixteen types in a popular book by Myers and

Briggs. Note that all three of these astronomer types are NTs. NT people are "Rationals" (or Conceptualizers). N stands for intuitive; T stands for thinking. People of the other broad temperaments, who aren't astronomers, are "Idealists," "Guardians," and "Artisans."

Of the three NT types, astronomers are more often INTJs. INTJs believe that competence plus independence equals perfection. Some of the occupations besides astronomy that are favored by INTJs are management consultant, economist, scientist, computer systems analyst, and university teacher (per Tieger).

According to Kiersey and Bates, NTs believe that to understand and control nature is to possess power. It is that desire for power that sets the NT apart from others. Power fascinates the NT. Not power over people, but power over nature. They want to be able to understand, control, predict, and explain realities. Scratch an NT, find a scientist.

NT characteristics: What NTs want is knowledge, to be competent, and to achieve. They seek to understand how the world and things in it work. Theory oriented. They trust logic and reason. Want to have a rationale for everything. Skeptical. Hunger for precision, especially in thought and language. Skilled at long-range planning, inventing, designing and defining. Generally calm. Foster individualism. Frequently gravitate toward technology and the sciences (per University Associates).

Common strengths of an NT person: Sees possibilities, works with the complicated, works out new ideas, is logical and analytical, organized, just, stands firm.

Common weaknesses of NTs: Inattentive to detail, impatient with the tedious, jump to conclusions, do not notice people's feelings, often fail to follow through

If you want to get a job working as an astronomer:
From: AOL, Careers and Work
Salary range in 2002 $78,686-$92,260
Minimum education Doctoral Degree
Job openings On the Decline
Availability of part time jobs Very Low

References:
For "More about Stephen Hawking"
1. Hawking, Stephen. "A Brief History of Time." Bantam Books, N.Y. 1988
2. Hawking, Stephen. "The Universe in a Nutshell." Bantam Books, N.Y. 2001
3. Hawking, Stephen. "A Briefer History of Time." Paperback, 2005
4. http://en/wikipedia.org/wiki/Stephen_Hawking

For "Competition Among Astronomers"
5. "NASA Kepler Telescope Will Search for New Earths." softpedia.com/news/NASA-039-s-kepler.
6. "How Long Until We Find a Second Earth?" discovermagazine.com/2008/nov/10
7. "Found 20 Light Years Away: the New Earth". dailymail.co.uk/sciencetech/article--450467
8. "Competition" Encyclopedia of Psychology. findarticles.com/p/articles/mi_g2699

For "What Kind of Person Likes Astronomy?"
9. Behrens, L. V., Telos Publications.
10. Tieger & Tieger, Little, Brown (page 195).
11. Kiersey and Bates, Prometheus Nemesis Books (page 47).
12. "1980 Handbook for Group Facilitators," University Associates, page 98.

# Chapter 9
# Old-time Astronomers

## Outstanding Astronomers of Long Ago

Aristarchus. Greek, born 310 BC. He was the first to say that the Earth revolves around the Sun. Archimedes wrote down Aristarchus' ideas.

Eratosthenes. Greek, born 276 BC. Found a way of calculating the size of the Earth. He measured the angles of shadows cast by vertical posts in Alexandria, Egypt and a town farther south. He calculated 28,500 miles circumference, which is close to the actual 24,900 miles.

Hipparchus. Greek, born 180 BC. Discovered the precession of the equinoxes. After studying earlier star observations, he noticed that the stars shifted eastward. He explained this shift as a slow forward motion of the vernal and autumnal equinoxes.

Ptolemy. Greek, born 100 AD. He said that the Earth was at the center of the universe and was motionless. He thought that everything in the universe moves either toward or around the Earth's center, at various rates of speed. Wrong. Nevertheless, he was a great man. He made a catalog of the locations and magnitudes of 1022 stars.

Copernicus. Polish, born 1473. He said that every planet in our solar system, including the Earth, revolves around the Sun, and that the Earth also spins on its axis once every day (which had made it appear to Ptolemy that the stars were moving around the Earth).

Brahe. Danish, born 1546. Telescopes were not invented yet, but he made and recorded observations that were much more accurate than had ever been made before, using only his eyesight and primitive instruments. However, he didn't believe Copernicus' theory.

Galileo. Italian, born 1564. Improved refracting telescopes and made the first effective use of them in astronomy. Discovered four Moons circling Jupiter. Discovered the laws of falling bodies and pendulums. Got in trouble with the church for believing Copernicus' theory.

Kepler. German, born 1571. Discovered three important laws of planetary motion:

Huygens. Dutch, born 1629. He said that light consists of a series of waves, and used that theory to study refraction of light. Newton said light consists of corpuscles. Both theories are correct.

All of these great men from before 1700 were wrong about some things but right about important others.

---

## The Odd Couple Who Couldn't Get Along

A fictitious conversation between Johannes Kepler and Tycho Brahe:

Kepler: "Don't ask me to look at stars, Tycho. My eyesight is terrible. I'm interested in pure theory. Nevertheless, I do need the facts that you're collecting."

Brahe: "I look because I want to be in touch with reality. You do your thing and I'll do mine."

Brahe and Kepler met in 1600, but their personalities clashed so much that they couldn't work well together, although Brahe was Kepler's employer. Brahe wouldn't even give Kepler all of his data about stars and planets.

My guess is that Brahe was a Myers-Briggs temperament type INTP (1% of the population), and that Kepler was a type ENTJ (5% of the population). Note that both men had _NT_ in their types (as almost all astronomers do), which means they are imaginative and logical, and crave knowledge.

Tycho Brahe (pronounced teekoh braw-hee) was an eccentric Danish aristocrat. A childless uncle kidnapped him from his parents, raised him, and sent him to law school. Tycho's nose was cut off in a duel and he used a silver piece to replace it.

Brahe wanted to be an astronomer. In 1572 he started studying a brilliant "new" star, which we now know to have been a supernova, in the constellation Cassiopeia. He watched it for two years from widely spaced places in Europe, and not seeing any parallax against other Cassiopeia stars, he concluded that it was outside our solar system. That meant that stellar objects outside our solar system could change, which was contrary to conventional wisdom, and it made Brahe famous. He later meticulously recorded the times and locations of many celestial objects from 1576 to 1591, using instruments of his own design.

Kepler was a German astronomer and mathematician. He knew that Copernicus believed the planets went around the Sun in basically circular orbits, with some added fudge factors called "eccentrics" and "equants." At first, Kepler accepted Copernicus' model of the solar system.

When Tycho Brahe was dying, in 1601, he relented and let Kepler use the rest of his unpublished observations. Kepler took over Brahe's job and data and studied the orbit of Mars for years, assuming at first that it had a Copernican circular orbit. Kepler knew that most of Brahe's observations, although made without telescopes, were accurate within 1 arcminute, so Kepler couldn't tolerate an apparent disagreement between Brahe's reliable historical data and the theoretical circular model. Also, Kepler knew there must be some force that keeps the planets from flying away from the Sun. Kepler finally came up with an idea that the Sun must have a strong magnetic field (wrong) that would let Mars have an elliptical orbit (right).

Kepler correctly described the elliptical motions of the six known planets in 1609 and 1619. It wasn't until 1687 that Isaac Newton explained why the planets move that way; it wasn't due to a magnetic field, but because of gravitational attraction. Kepler's laws were then touched up very slightly to use the collective center of mass of the Sun and a planet, instead of the center of the Sun alone.

Kepler's original famous laws, which apply even to modern satellites, are:

1. Each planet travels in an elliptical orbit. The Sun is at one of the two foci of that ellipse.
2. The way the planets revolve around the Sun is that an imaginary line that connects planet and Sun sweeps out equal areas in equal times. A planet therefore moves slower when it is farther from the Sun.
3. The squares of the sidereal periods of any two planets are proportional to the cubes of their mean distances from the Sun. For any planet: The time

to go around the Sun, squared = Distance from the Sun, cubed.

Kepler's third law enables astronomers to calculate the relative distances of the planets from the Sun by merely using measurements of how long it takes each of the planets to go once around the Sun.

---

## Giants Of Astronomy

Ptolemy (100-170 AD). A remarkable astronomer. He invented the astrolabe, an instrument with which the positions of the stars could be used to determine an observer's position on land or sea provided that he knows the date. It was the best available navigation instrument for centuries. He also described a model of the universe with the Earth at the center, that was accepted and refined for more than a thousand years, but which turned out to be incorrect.

Copernicus (1473-1543). A brilliant Polish physician, he declared that the Earth was not the center of the universe as Ptolemy had said, but that the Earth revolved around the Sun. However, Copernicus thought that the Sun was the center of the entire universe. The church rejected his idea that the universe was not centered on humanity, but Copernicus escaped punishment because he delayed publishing his hypothesis until the year of his death.

Galileo (1564-1642). This great scientist built a refracting telescope in Italy for use in astronomy and used it to gather new information about the planets. He also worked on other scientific subjects, including gravity and the pendulum.

Kepler (1571-1630). A German astronomer, he used mathematics to formulate three laws of planetary motion. (a) The planets' paths are elliptical. (b) A radius vector of a planet's orbit sweeps out equal areas in equal lengths of time. (c) The square of a planet's year is proportional to the cube of its average distance from the Sun. Copernicus had thought that planetary orbits were circular.

Newton (1642-1727). He invented the reflecting telescope, for which it was practical to have a large aperture, and he made many other marvelous scientific discoveries, including his "laws" of mechanics. (a) A body at rest or in motion remains at rest or in motion unless acted upon by a force. (b) Acceleration of a body equals the force applied to the body divided by the mass of the body. (c) A force has an equal and opposite reaction.

Einstein (1879-1955). In his Special Theory of Relativity published in 1905 and his General Theory of Relativity in 1915, he proposed radical ideas about the nature and interrelationships of matter, light, space and time, for example, that mass and energy are interchangeable, space is curved by the presence of mass, and time isn't the same everywhere. Experiments later showed most of his theories to be true.

Hubble (1889-1953). American astronomer Edwin Hubble proved that the other galaxies are independent of the Milky Way. By 1930 he had confirmed a concept, proposed by others, that the universe is expanding. (Recently, astronomers have started to believe that the rate of this expansion is increasing.)

There's an old saying that ontogeny recapitulates phylogeny, which means that steps by which an embryo develops into an adult are similar to the steps in the entire evolutionary history of a species, such as humankind. Here's a new saying:

Astronomy recapitulates ontogeny. A baby thinks he is the center of the universe, just as most people thought that humans were the center of the universe in Ptolemy's day. Step-by-step, the giants of history have raised the infant astronomy to grow up into a sophisticated and confident adult.

## A Famous Church Controversy

"The Roman Catholic Church gave more financial and social support to the study of astronomy for over six centuries, from...the late Middle Ages into the Enlightenment, than... probably, all other institutions." That's the first sentence in the book "The Sun in the Church" by J. L. Heilbron. In the next 365 pages the author builds a pretty good case for that statement by presenting a lot of detailed scientific examples.

One of the reasons the church was so interested in astronomy was the problem of setting the time of Easter. We start with the full Moon that occurs on the vernal equinox or, if none there, the next full Moon after the vernal equinox. The first Sunday after that full Moon is Easter. Although the equinox, the full Moon, and Sunday occur on different days at different longitudes on the Earth, the church leaders wanted Easter to be celebrated on the same day in all places. They had a standing committee that studied this problem for hundreds of years.

J. L. Heilbron, from the University of California, includes a section entitled "The Matter of Galileo," relating how Galileo got in trouble with the church. Historians know that Galileo was sentenced to house arrest for life because he advocated Copernicus' theory that the Earth revolves around the Sun.

Heilbron says Galileo's problem with the church was due mainly to his obnoxious overreaching and uncompromising attitude. He had low emotional intelligence. Later, some Italian clerics actually taught and advanced the Copernican theory, but with "diplomatic discretion." Heilbron avers that some of those clerics were skilled mathematicians and very good astronomical observers. For centuries there were many scholarly "niches" that were protected and financed by the Catholic Church, and in which science and mathematics thrived. He feels that modern historians should mention those facts.

So this famous controversy keeps on percolating.

## Partly Right

What some great astronomers could have said to earlier great astronomers:

Claudius Ptolomy, 100 – 170 AD. "The planets and stars circle around the Earth as a center. They ride around as though they are on huge transparent spheres."

Copernicus, 1473–1543. "Wait a minute, Claudius. I can see why people have believed for the 1300 years since you died that the stars orbit the Earth, but they don't, although for remote stars it's hard to tell that it's wrong. However, it's a bigger stretch to say the Sun and planets go around the Earth. The Earth and other planets in our solar system actually go in circles around the Sun."

Kepler, 1571–1630. "That's partly right, Nicholas, but I've been studying the planet locations measured by my cohort

Tycho Brahe, and they show that the orbits of the planets are not circular but elliptical."

Newton, 1642–1727. "That's right, Johann, and the planets go around mainly under the influence of the Sun. They feel a gravitational attraction for the Sun."

Einstein 1879–1955. "Nice going, Isaac, but it isn't gravitational attraction that keeps the planets from flying away from the Sun. It's a curvature of space near the large mass of the Sun. But at least the universe is remaining the same size."

Hubble, 1889–1953. "Good, Albert, but that last part is wrong. My observations confirm a few other people's ideas that the universe is expanding. Also, I find that the farther a star is from Earth the faster it is moving away. The radial speed of each star is proportional to its distance from us."

Gamow, 1904–1968. "I think you're right, Edwin. And the fact that the universe is expanding shows that there must have been a tremendous Big Bang about 14 billion years ago."

Prof. George Efstathiou of the University of Cambridge, in 2002: "Yes, it is expanding, George and Edwin, and it appears to be expanding faster and faster. In 1999, someone said the brightness of supernovae explosions in remote galaxies appeared to require that the universe be filled with a strange kind of dark energy that causes its expansion to accelerate. In 2002 our team of 26 British and Australian scientists reported a very clever new test for acceleration of expansion. We compared the structure of the clustering pattern of galaxies at two very different times: (a) Now, i.e., billions of years after the Big Bang, and (b) at a time when the universe was only 300,000 years old. The pattern 300,000 years after the Big Bang is shown by cosmic microwave background radiation, which preserves the pattern from long ago. Then we used geometry to draw pictures of the universe. Expansion is speeding up, into the infinite future."

Only partly right? Probably. But we've made great strides in the last 2000 years by being only partly right. Progress has been especially fantastic in the last 400 years, because our ability to analyze is speeding up too!

# Chapter 10
# Poetry About Astronomy

## Shakespeare

William Shakespeare (1564-1616), referred to the Moon one hundred and seventy times in his writings.

*Romeo and Juliet,* act II, scene ii
"But soft! What light through yonder window breaks.
It is the east, and Juliet is the Sun!
Arise, fair Sun, and kill the envious Moon,
Who is already sick and pale with grief,
That thou her maid art far more fair than she."

The Moon doesn't have any clouds, which is one reason it reflects only a small percentage of the Sunlight it receives.

*Romeo and Juliet, act II, scene ii*
"*Romeo*: Lady, by yonder blessed Moon I swear
That tips with silver all these fruit-tree tops
*Juliet*: O! swear not by the Moon, the inconstant Moon,
That monthly changes in her circled orb,
Lest that thy love prove likewise variable."

The Moon's schedule must have been very confusing to early observers because it is affected by so many factors. A lunar calendar year is twelve full cycles of the Moon, e.g., twelve new Moons, and a lunar year is only 354-1/3 days. But a solar calendar year, which is the time it takes for the Earth to revolve once around the Sun, is 365-1/4 days.

*Hamlet, act II, scene ii*

"Doubt thou the stars are fire,
Doubt that the Sun doth move;
Doubt truth to be a liar;
But never doubt I love."

Although it is figurative to say that the stars are "fire,' in a way it is true. They are aggregations of extremely hot material, mostly hydrogen and helium. In the constellation Scorpius, the binary pair Antares A (which is red) and the smaller Antares B (which is blue) have surface temperatures of about 3,000° Kelvin and 15,000° Kelvin respectively.

*A Midsummer–Night's Dream, act II, scene i*

"Since once I sat upon a promontory,
And heard a mermaid on a dolphin's back
Uttering such dulcet and harmonious breath,
That the rude sea grew civil at her song,
And certain stars shot madly from their spheres
To hear the sea-maid's music."

For 1300 years before Copernicus the stars were thought to be on spheres that rotated in the heavens with Earth at the center. Some celestial bodies such as comets and meteors that burn themselves out in the Earth's atmosphere were not in fixed positions, but "shot madly from their spheres."

*Julius Caesar,* act III, scene

"I am constant as the northern star,
Of whose true-fix'd and resting quality
There is no fellow in the firmament."

That's what Caesar said in rejecting three warnings not to go to the senate one day, and it was his undoing.

*King Henry the Fourth, Part I, act I, scene ii*

"Diana's foresters, gentlemen of the shade, minions of the Moon."

These are euphemisms for highway robbers. The Romans called their Moon goddess Diana, lover of the woods and the wild chase.

---

# FitzGerald and Frost

Edward FitzGerald The Rubaiyat of Omar Khayyam

"Wake! For the Sun who scatter'd into flight
The Stars before him from the Field of night,
Drives Night along with them from Heav'n and strikes
The Sultan's Turret with a Shaft of Light."

It is the scattering of sunlight by the Earth's atmosphere that makes stars invisible from the Earth (but not from satellites), in the daytime. Very small particles in the atmosphere (less than one wavelength) scatter the blue light much more strongly than the red light, making the whole sky appear blue. This is Rayleigh scattering.

Robert Frost *Bravado*

"Have I not walked without an upward look
Of caution under stars that very well
Might not have missed me when they shot and fell?
It was a risk I had to take–and took."

Frost was being playful, knowing that the probability of his being struck was negligible. Meteors are streaks of light caused by meteoroids (shooting stars), which are small pieces of dust that burn up in our upper atmosphere because of friction with the air. Larger bodies that make it all the way to the Earth are called meteorites. About 3300 meteorites hit the Earth each year. Most of them are unrecorded because they fall in oceans, deserts, etc.

Anonymous Aztec Indian The Flight of Quetzalcoatl.
"It ended
With his body changed to light,
A star that burns forever in that sky."

I don't know whether all deceased Aztec Indians are believed to become stars in the sky, or whether only the gods such as Quetzalcoatl are thought to become stars. This may be an Aztec version of heaven.

---

## Emily Dickinson and Others

Here is poet Emily Dickinson's "take" on a daily astronomical phenomenon.

I'll Tell You How the Sun Rose
I'll tell you how the Sun rose,
A ribbon at a time.
The steeples swam in amethyst,
The news like squirrels ran.

The hills untied their bonnets,
The bobolinks begun.
Then I said softly to myself,
"That must have been the Sun!"

But how he set, I know not.
There seemed a purple stile

Which little yellow boys and girls
Were climbing all the while

Till when they reached the other side,
A dominie in gray
Put gently up the evening bars,
And led the flock away.

This Dickinson poem is based on the painting "Soleil Levant" by Monet, which has some amethyst color. She describes the Sunset differently from the Sunrise, although, from photographs, astronomers can't tell the difference.

(A dominie is a clergyman.)

Meleager *The Greek Anthology*

"Farewell, Morning Star, herald of dawn,
and quickly come as the Evening Star,
bringing again in secret
her whom thou takest away."

Venus is the Morning Star at certain times and the Evening Star at others. This is because it is not very far away from the Sun, so it is sometimes seen trailing the Sun across the sky and sometimes leading it.

Walter De La Mare_ *The Wanderers*

"Wide are the meadows of night,
And daisies are shining there,
Tossing their lovely dews,
Lustrous and fair;
And through these sweet fields go,
Wanderers amid the stars–
Venus, Mercury, Uranus, Neptune, Saturn, Jupiter, Mars."

Although their motions are now well understood, planets seemed to wander against a backdrop of the more regular constellations of stars.

James Rado and Gerome Ragni *Hair, 1966*
"When the Moon is in the seventh house
And Jupiter aligns with Mars,
Then peace will guide the planets,
And love will steer the stars.
This is the dawning of the Age of Aquarius,
The Age of Aquarius."

The signs of the zodiac are stellar constellations that are distributed throughout a belt of the sky about 16 degrees wide, centered on the ecliptic. Different zodiacal signs are visible at different times of the year.

# Chapter 11
# Motions of the Earth

## The Earth Moved When Erik Moved

Eric Hall, an amateur astronomer, moved this summer. He went 12 miles from Olmsted Falls, Ohio to a house in Parma. From noon to midnight during that same move, he also moved around a half circle because of the spinning of the Earth on its axis. So at midnight, at Ohio's latitude, he was about 5,500 miles farther away from the Sun.

Since the Earth's axis is cockeyed, Eric also traveled south about 2,400 miles closer to the plane of the ecliptic.

Besides, the Earth scooted about 800,000 miles in its orbit around the Sun during that interval. All this is making me dizzy.

By the way, our Galaxy revolves in space around its center, so our entire Solar system moved about 6 million miles through space in that half day.

I should mention that, because of expansion of the universe, Erik moved about 30 million miles away from a galactic cluster in the nearby constellation Virgo.

Yep, Erik sure moved that day!

# Precession of the Earth's Axis

While a wedding, graduation or funeral may have a procession, the Earth has something more interesting–precession. Its axis precesses westward around the ecliptic with a period of 25,800 years.

Precession is a slow change in the direction of the axis of rotation due to the application of an external torque. A gyroscope is a good example. The Earth has an equatorial bulge, and it is mainly the gravitational attractions of the Sun and Moon on that midriff that cause the Earth's axis to precess. They pull on the bulge in a direction to press the equatorial plane toward the plane of the ecliptic. There is also a small pull from the planets. Together, those torques currently cause a precession of about 50 arc seconds each year.

It's hard to believe, but precession was first described almost 2200 years ago by Hipparchus!

During one full period of precession, the celestial poles of the Earth trace out circles 23.4 degrees in radius on the celestial sphere. The equinoxes precess, consequently the stellar coordinates do also. Coordinates such as right ascension that refer to an equinox as their zero point change with time.

That's why star catalogs give positions for a "standard epoch," which is currently the start of the year 2000. The standard epoch is an agreed time, which is changed about every 50 years. Because of the precessional changes in stellar coordinates, the nominal positions of the celestial poles and the two pole stars change.

# Cause of the Precession of Earth's Axis

In the last 5000 years, the position of the Sun at the vernal equinox (first day of spring) has moved out of the constellation Taurus through Aries and into Pisces. The position of the north celestial pole has also changed since 3000 BCE; it has moved from the unfamiliar star Thuban (in Draco) to a place very close to Polaris.

The Earth spins on its axis like a gyroscope, with its equator inclined 23.4 degrees from the plane of the Earth's orbit (the ecliptic). When torque is applied to deflect the axis of a gyroscope the gyroscope precesses.

Precession is a slow periodic change in the direction of the axis of rotation, caused by an external torque. According to Newton's laws, a body in motion remains in motion unless acted upon by an external force. Similarly a body that is rotating continues to rotate unless acted upon by an external torque.

In the case of a gyroscope, when a torque is applied that tends to tip the axis of rotation, the body reacts by precessing, thereby conserving angular momentum. If a torque on the axis of a gyroscope is constant, the axis twists around in a circle, and that is what happens with the Earth. Its axis traces out cones near the north and south poles.

The torques that cause the precession come from the Moon, the Sun and the planets, all of which tug on the Earth by gravitational attraction. The Moon and planets are approximately in the ecliptic plane, so the Moon, Sun and planets are almost always above or below the equatorial plane of the Earth.

Our Earth is not quite spherical; it bulges at the equator and is slightly flatter at the poles, so it experiences a torque from the Moon and the Sun due to their greater pull on the equatorial bulge. Torque on the bulge on the side of the Earth near the Moon is in an opposite direction from torque on the bulge on the far side, but because the near side is 8000 miles closer to the Moon, the near-side torque is more powerful and the torques don't cancel.

The Sun is only 37% as effective as the Moon in causing our precession because the Moon is only 240,000 miles away, while the Sun is 93 million miles away. The torques that the Moon and Sun apply are in a direction that tends to tip the equator toward the ecliptic plane; that is they tend to press the axis of rotation of the Earth toward becoming perpendicular to the ecliptic. But instead of tipping, the Earth reacts by precessing. The north pole and south pole creep around in circles of 23.4 degree radius, taking 28,500 years to go full circle.

Precession causes the equinoxes (where the equator intersects the ecliptic) to migrate in a slow westward motion around the ecliptic, moving earlier by about 50.28 arc seconds per year. The stars also give the appearance of drifting by that amount. To define star positions unambiguously, astronomers have had to adopt a "standard epoch," which by international agreement is the start of the year 2000.

In order to prevent the seasons from slowly marching around the calendar, our calendars use a "tropical year" instead of a "sidereal year," so we'll never have winter in July.

## Just a Second–Earth's Rotation is Slowing

The Earth is rotating slower and slower. As a result we have seen earthquakes from tectonic plate movements, volcanic eruptions, and the Indonesian tsunami of a few years ago, in which tens of thousands of people were killed.

These things happen because there is a war on Earth, of gravitational force versus centrifugal force. Gravity tries to make the Earth a perfect sphere and centrifugal force (from Earth's daily rotation) tries to pull the area around the equator to protrude, leaving flat poles.

When the Earth slows down the centrifugal force diminishes and lets gravity draw the Earth closer to a perfect sphere, with a smaller "spare tire" at the equator. The crust of the Earth shrinks, just as a shrinking plum develops a lot of prune wrinkles.

Disruptions like those wrinkles occur in the Earth's crust as the crust becomes more spherical. Continents move around. Some spaced-apart tectonic plates move to abut each other and buckle upwards to create mountains and undersea ridges, and overlapping plates slide farther over each other.

Important causes of the slowing are the tides that the Moon produces on the Earth. Of course the Earth is rotating beneath the tides, so the tides move to follow the Moon as the Earth turns. Tidal bulges are not centered on an imaginary line from the center of the Earth to the center of the Moon because of frictional drag of the rotating Earth on the water, at the sea floor. The tidal bulges are dragged along so as to be leading (east of) that imaginary line.

Because of the leading easterly location of the Earth tides, the gravitational attraction of the Moon acting on the tides produces torque that slows down the Earth. The tides are like brake shoes that apply brakes to the rotating Earth.

The gradient of the Moon's gravity field creates a tide on the far side of the Earth as well as on the near side, because the Moon's gravity field is stronger on the side of Earth nearer the Moon than on the far side. The Moon's gravity gradient stretches the Earth (both its solid and watery parts), to produce the tides on both sides. Centrifugal force due to the Earth's orbit around the Sun and our gravitational attraction for the Sun tend to hold the Earth in a nearly circular orbit while the Moon does this stretching of the Earth and its oceans.

There are several types of slowing. (a) Secular slowing, of about 1 to 2.3 ms per day per century. (b) Irregular slowing, which may increase for 5 or 10 years, then decrease, with a net accumulation of an amazing 44 seconds since 1900 (probably due to magneto-hydrodynamic effects of the rotation of the liquid iron core). (c) Periodic slowing, with periods of one year and of six months. The periodic one-year slowing is due to seasonal variations in winds which make the Earth late about 30 ms on June 1 and ahead about 30 ms on Oct. 1. The six-month variations are caused by tidal action of the Sun on the Earth. The Sun-caused tides are only half as great because tidal effects vary inversely as the cube of the distance and the Sun is so much farther away than the Moon.

Nowadays the period of Earth's rotations is measured with cesium-beam atomic clocks, and compared with astronomical time based on the orbital motion of the Moon about Earth and of the planets about the Sun. Cesium clocks, which were invented about 1955, are based on a theoretical idea by Columbia University physicist Isidor Rabi. Cesium-133 is an isotope of a metal, which is heated and processed to be a gaseous beam. All of its electron orbits are very stable except the one farthest out. The outermost electron shifts out and back with an inherent resonant frequency when it is synchronized with a local oscillator tuned to that frequency. Electronic dividers keep track of the time.

We just reset our clocks by 1 second, at the end of 2005. People who opposed resetting the clocks said that computers don't know how to handle a 61-second minute, but those in favor of resetting them reminded us that, a few years ago, the anticipated Y2K computer problems were easily overcome, and that it would be helpful to have the clocks match astronomical observations.

The Moon used to be gradually slowing down also, centuries ago, by tides on the semi-plastic Moon caused by the Earth. Those tides slowed the rotation of the Moon until it rotated only once for each revolution of the Earth around the Sun. Thereupon the tidal torque on the Moon became zero, so the Moon started always showing the same face to the Earth, as it still does.

About 900 million years ago Earth's day was only about 18 hours long. Eventually the Earth will slow down so much that it will continuously present the same face to the Moon, whereupon the Moon-caused slowing torque will stop.

---

## Earth's Peripatetic Magnetic Poles

Derek (Pragmatic amateur astronomer): What do you know about the magnetic poles? I know that James Ross located the magnetic north pole in 1831.

Anne (Theoretical amateur astronomer): Yes. Then his ship was stuck in the ice for four years. Seventy three years later Roald Amundsen went up there and found that the north pole had moved about 30 miles.

Derek: I read that the north pole, which is presently in Canada, appears to be migrating from Canada to Siberia. It had been moving northward about six miles per year, but it sped up to about 25 miles per year recently. It has moved 685 miles in the last 150 years.

Anne: Larry Newitt of the Geological Survey of Canada flies up there periodically to keep track of its location. Amateur astronomers can compare a careful magnetic compass reading with the catalog position of the North Star, which isn't quite true celestial north, and keep track of the pole's movement over the years. The south pole of the permanent magnet needle of a compass seeks the north. You know, Derek, opposites attract.

Derek: Uh huh. They do, but it's only temporary. They don't stay together like people of similar interests do.

Anne: They can stay together, provided both of them give up the Pygmalion project of trying to change the other so they match. That would destroy the "opposites attraction," Derek.

Derek: Really? Hmmm. Never thought of that!

Anne: Anyhow, an imaginary line between the magnetic north and south poles is tilted about 11 degrees from the Earth's axis of rotation, with the south pole being four times as far away from that axis as the north pole is.

Derek: To me, a diagram of the Earth's magnetic field looks like the field of a permanent magnet, which of course is produced by coordinated directions of electron motions in iron atoms, but the Earth's field may not be produced the same way.

Anne: It isn't. The temperature of iron in the core of the Earth is much greater than iron's 1043 K Curie temperature, at which atomic electrons start to move at random. Rather, most scientists think the Earth's magnetic field is created by a dynamo effect, that is, by electric currents in the outside of the molten iron core, which is rotating.

Derek: The core is very chaotic, and the theories of magnetohydrodynamics are also dauntingly complicated.

Anne: If the molten core acquires an electrical charge, say by friction between its liquid layers, the overall rotation of the core can carry those free charges around to produce a current loop, which creates an electromagnetic field.

Derek: I do think scientists are pretty sure our magnetic field is connected with the rotation of the Earth, because even though Venus has a similar iron core, it does not have any measurable magnetic field. That's probably because its rotation is just too slow. It has a 243 Earth-day rotation period.

Anne: Yes. You know, some folks think our magnetic field is healthful. In the time of the dinosaurs the magnetic field was 80% stronger than now, which may account for the fact that only much smaller animals can exist on Earth now, the largest being elephants.

Derk: Theory says that the magnetosphere, the magnetic field that surrounds the Earth, traps charged particles in areas called the Van Allen belts. The magnetosphere also protects the Earth by deflecting the solar wind, which is a stream of ionized gases coming from the Sun.

Anne: Geophysicists piece together the magnetic history of Earth by studying the permanent magnetism of ancient rocks. In Hawaii there are layers of iron-rich lava rocks that have retained their original magnetism from when they cooled and solidified. Carbon dating shows how old each rock is. The rocks' magnetic fields are of various strengths and they point in various directions, all of which goes in the data book.

Derk: Yeah, and from this history, reversals of magnetic polarity have been found to have occurred about 171 times in the last 71 million years, and very irregularly. After a reversal, the north pole appears in the south and vice versa. That happens on average about 250,000 years apart, although the most recent reversal was more twice that long ago, and it could be millions of years. The next one is expected by about 3000–4000 AD. A reversal takes several thousand years to complete. Incidentally, I know the Sun reverses its magnetic field every 11 years, when the Sunspot cycle is at a maximum, which will be in 2012.

Anne: Just before a reversal, Earth's magnetic field strength has usually become very weak. If that happens next time, the magnetosphere might fail to protect the Earth from excessive solar radiation, resulting in a lot of cancer. Migrating birds might lose their sense of direction, and it's conceivable that cosmic rays might even kill most of the creatures on Earth.

Derek: People who predict an early end of the world tend to focus on that kind of information.

Anne: More conventional scientists think a reduced magnetic field would not be very bad, because our magnetic field probably wouldn't vanish completely. It might just get more tangled, such as by having extra poles, like a north pole that pops up in India or a south pole in the Pacific Ocean. According to that theory we'd still be protected from the Sun by the magnetosphere.

Derek: I sure hope they're right, and that you're right about the Pygmalion project, Anne.

References:

For " Precession of the Earth's Axis"

1. Zeilik, Michael: "Astronomy," John Wiley & Sons, Inc., New York. 7th ed., 1994.
2. Chaisson, Eric and McMillan, Steve: "Astronomy Today," Prentice-Hall, 3d ed. 1993.
3. Illingworth, Valerie, Editor.. "Facts on File Dictionary of Astronomy" Market House Books Ltd., 3d ed., 1994.

# Chapter 12
# Cosmology

## The Mystery of Dark Energy

For most of the first five billion years after the Big Bang, the universe was expanding at a slightly decelerating rate. Gravity was slowing the expansion. Then at perhaps nine billion years ago (controversial) the expansion started accelerating. What is causing that? Dark energy is strongly suspected, but no one is sure what dark energy is. This is like a whodunit in which they can't find a body and the only evidence is circumstantial. Here are some definitions of the phenomena involved.

Dark Energy

A form of energy thought to permeate all of space and to increase the rate of expansion of the universe.[1] About 74% of the mass-energy of the universe may be dark energy, 22% may be dark matter, and 4% ordinary matter. The effect of dark energy is the opposite of gravitational attraction. It causes a repulsion of other things, and accelerates expansion of the universe. Since energy and mass are related by $E = mc^2$, Einstein's theory of general relativity predicts that dark energy will have a gravitational effect.

Supernovae

Observations of many supernovae provide the most direct evidence of dark energy because collectively they show acceleration of the expansion of the universe. An ordinary nova is a binary star system in which there is a sudden and unpredictable increase in brightness of about ten magnitudes. One magnitude is a brightness ratio of about 2.5. A supernova is a star that may temporarily be even a hundred times brighter than an ordinary nova. It runs away in a violent thermonuclear explosion during a few days or months.

The redshift of light from a supernova shows how fast it was receding from us at the time it emitted the light. But it is also necessary to know its distance from Earth. To determine the distance, astronomers measure the observed brightness on Earth of a type 1a supernova, and, because the theory of explosions of type 1a supernovae is known, they already know this type's absolute magnitude (actual brightness). The distance from Earth of a supernova can be determined from the ratio of its observed brightness to the theoretical absolute magnitude. The redshifts and the distances of many supernovae are used to plot the expansion history of the universe.

Cosmological Constant

A constant term that Einstein arbitrarily added to his field equations of general relativity theory in 1915 to make the universe look static (as most people then believed it was), instead of expanding or contracting. The cosmological constant (designated by the symbol lambda, like an upside-down V), represents a constant energy density that fills all of vacant space. The discovery in the 1920s that the universe is expanding appeared to make the cosmological constant zero, and therefore unnecessary. However, starting in 1998, evidence of acceleration indicated that the cosmological constant may have a small but significant positive value–enough to cause the acceleration. It can be small and still be very effective because it is everywhere in otherwise empty space. It is sometimes called "vacuum energy."

A possible alternative to the cosmological constant is a form of dark energy dubbed "quintessence," which is a field whose energy density can vary in time and space. Quintessence is the most popular of several alternatives to the cosmological-constant theory.

Dark Matter

Physical objects or particles that emit little or no detectable radiation of their own and are believed to exist because of evidence of otherwise unexplained gravitational forces observed on other astronomical objects.

---

## Quotations about Dark Energy

Here are brief excerpts from fourteen Internet articles.

1. "Dark energy entered the astronomical scene in 1998, after two groups of astronomers made a survey of exploding stars (supernovas), in a number of distant galaxies. These researchers found that the supernovas were dimmer than previously believed, and that meant they were farther away than previously believed. The only way for that to happen, the astronomers realized, was if the expansion of the universe had sped up at some time in the past."
2. "The last thing the two teams expected to find was that the expansion of the universe is not slowing at all. Instead, it is accelerating."
3. "...the chance discovery of the most distant supernova has revived a discarded theory of Albert Einstein... Astronomers using the Hubble Space Telescope found the exploding star about 10 billion light-years from Earth."

4. "Under Newton's laws, stars and other heavenly bodies pull on one another through the force of gravity. A countervailing propulsion, like a big explosion, could overcome that attraction, but once it fizzled out, gravity would start pulling things together again. Either way, matter in the universe should be moving–either hustling out into space or clumping into a kind of cosmic hairball.

But the universe that Newton and Einstein knew was a tame, stable place. …even a theory as powerful as relativity failed to explain it. So Einstein added an arbitrary term to his equations. Mathematically, it acted like a repulsive force spread smoothly throughout the universe. Where gravity pulled, he said, this force pushed back in equal measure."

5. "…Einstein conjectured that even the emptiest possible space, devoid of matter and radiation, might still have a dark energy. When Edwin Hubble discovered the expansion of the universe, Einstein rejected his own idea, calling it his "greatest blunder." "Through recent measurements of the expansion of the universe, astronomers have discovered that…some form of dark energy does indeed appear to dominate…and its weird "repulsive gravity" is pulling the universe apart."
6. "Observations of distant supernova have suggested… that dark energy dominates the universe…"
7. "…dark energy pervades every nook and cranny…"
8. "Using the world's most powerful telescopes to make radio pictures of thousands of distant quasars–some of the brightest objects in the sky–the scientists calculated that two thirds of the cosmos is made up of dark energy. …astronomers have calculated that the total mass of all the visible galaxies only makes up about one-third of the

critical density needed to satisfy the best current theory about the early universe."

9. Critical density is "The factor that determines the dynamic behavior of the universe, i.e. whether it is open or closed."
10. "For the first time we have a plausible, complete accounting of matter and energy in the universe. Expressed as a fraction of the critical density, it goes like this: Neutrinos, between 0.3% and 15%; stars, 0.5%; baryons (total), 5%; matter (total), 40%; smooth dark energy, 60%, adding up to the critical density." (Some counted twice.)
11. "Baryonic matter is normal matter containing baryons, i.e. protons and neutrons. Dark matter has mass but is undetectable except by its gravitational effects. It is thought that a large proportion of dark matter could be composed of non-baryons such as WIMPs, which is an abbreviation for Weakly Interacting Massive Particles."
12. "The universe is made mostly of dark matter and dark energy, and we don't know what either of them is."
13. "To solve it, Berkley Lab physicists, astronomers... propose to launch a satellite named SNAP -- the Supernova/Acceleration Probe."
14. "This is how cosmologists summarize things: a static universe with no matter (if such a thing were possible) would have no curvature. If, however, the empty universe were expanding it would have negative overall curvature. Increase the mass density from zero and the curvature would be less negative. Add still more mass and you might reach a net zero curvature."

The most popular version of the Big Bang theory now is that the curvature of the universe is zero.

# Dark Matter

The Earth isn't dark matter; it reflects light that originates in the Sun. Dark matter is matter that doesn't emit or reflect or absorb visible light, radio waves, X rays, or any other kind of electromagnetic energy. Earth, on the other hand, is made of baryonic matter, which contains protons and neutrons, so it reflects.

Dark matter has mass but is detectable only by its gravitational effects. The gravitational effects include the motion of satellite galaxies (which are relatively small galaxies that rotate around a much larger galaxy), and the bending of light from distant galaxies by nearer galaxies. There may be thirty times as much dark-matter mass as visible matter in the universe. The mass of a galaxy seems to be much larger than the mass of its stars.

A lot of the dark matter in the universe may be non-baryons such as Weakly Interacting Massive Particles, called WIMPs. In the last few years, scientists have been searching for WIMPs by using sensors in underground laboratories in Minnesota, France, and Italy.

Astronomers are in the dark about dark matter (and dark energy). They want to learn the composition of dark matter to determine how the universe evolved. The total mass of all the visible galaxies adds up to only about one-third of the critical mass needed to satisfy the best current theory about the formation of the universe. Critical density is the amount of mass that determines whether the universe is open or closed.

In the last twenty years the cold dark matter theory has predicted the formation of galaxies having properties that are very similar to those of actual galaxies.

# Evidence of Dark Matter

Ouch! It's one sadistic blow after another for us humans. Not only has Darwin said we are descended from lower forms of animals. Not only has Copernicus said we are not the center of the solar system. Now astronomers say we aren't even made of the most popular materials.

Dark energy is believed to be about 74% of the mass-plus-energy total of the universe. Dark matter is believed to be about 22%. That leaves only 4% for baryonic matter such as protons and neutrons, of which we humans are made. And most of that 4% consists of stars and intergalactic gas.

Strong evidence of the existence of dark matter comes from galaxies, which tend to aggregate into clusters. Our galaxy, the Milky Way, is a member of a cluster called the Local Group. Clustering is caused by mutual gravitational attraction of the galaxies.

However, stars can orbit in their galaxies and a cluster of galaxies can rotate, so they also experience centrifugal force. The total of all the visible matter in a cluster doesn't provide enough gravitational attraction to prevent the galaxies, at their present high orbital speeds, from flying farther apart from each other.

But they don't fly apart, so there is a missing-mass problem. Calculations show that about 90% of the matter involved in galaxy clustering must be invisible. It is the ominous-sounding "dark matter."

In 1933, astrophysicist Fritz Zwicky said that the gravity of the visible galaxies in a cluster would be far too small to explain their rotational behavior; something more is required. About 1970 a young woman, Vera Rubin, accurately measured the velocities of edge-on spiral galaxies, and said that upwards of 50% of the mass of galaxies was contained in a relatively dark galactic "halo."

Besides the velocities of galaxies in clusters, subsequent evidence of dark matter includes (a) gravitational lensing of background objects by clusters of galaxies and (b) the temperature distribution of hot gas in and around galaxies.

The exact composition of dark matter isn't known, but it apparently consists of physical objects or particles that emit or reflect little or no detectable electromagnetic radiation. Those particles may include neutrinos, elementary particles such as WIMPs (weakly interacting massive particles), and axions, and there may also be some dwarf stars and planets and nonluminous gas. These furtive characters disclose their presence mainly by gravitational attraction on visible bodies.

By now, astronomers have garnered so much observational evidence that most agree that dark matter does exist.

---

## Redshift, Blueshift and What They Tell Us

Space is an equal-opportunity carrier, like a modern city bus. It moves all passengers, white, yellow, red, and black light at the same speed. Electromagnetic waves, which also include heat, radio waves, X-rays, gamma rays, etc.–all zip along at 3 x 108 meters per second in a vacuum. In other media the dielectric constant reduces the speed; e.g., the waves travel much slower in water.

Redshift is the conversion of electromagnetic radiation to longer wavelengths, and blueshift to shorter. Redshift is a misnomer if the pre-shift wavelength was longer than red light, but it is used anyway.

In astronomy redshift is usually denoted by z, which is the observed wavelength minus the emitted wavelength, all divided by the emitted wavelength. Z is therefore the per-unit change in wavelength.

When a pattern of emission or absorption lines is redshifted or blueshifted, the new pattern, which is measured by a spectroscope, is a scale model of the original pattern. Incidentally, Planck's constant says that a photon of longer-wavelength radiation has less energy.

Several basic phenomena, some of which are almost incredible, can cause redshift.

1. A phenomenon that is analogous to the Doppler effect that changes the pitch of sound waves. A distant star may be producing light of a certain color, say yellow. It launches a positive lobe of light onto free space, which starts traveling toward Earth. A half period later it emits a negative lobe, then another half-period later, a second positive lobe. By that time the first positive lobe has traveled for one period from where it was launched.

But during that one period the star has moved farther away from the Earth. Although the star sends the second positive lobe into space exactly one period later than the first positive lobe, the wavelength in space is greater than it would have been from a stationary source, because the star has been moving away. The star is actually launching a longer-than-yellow wavelength. An observer on Earth sees light of greater wavelength than would be measured from the same known elements on the Earth. That's the usual Doppler effect.

2. A second cause of Doppler effect is the expansion of space. Photons traveling through expanding space are stretched en route. This is called the "cosmological redshift." It is not due to the launching of longer wavelengths by the distant galaxies; instead, the travel route itself is stretching the waves.
3. Another mechanism that produces a redshift is the "relativistic Doppler" effect. When a source is moving at close to the speed of light, it exhibits what is called the Lorentz factor. Strangely, in this case the radiation is redshifted (or blueshifted), even if the motion of the source is perpendicular to the line of sight of the observer. Ordinarily, if a radiating object is approaching us, its radiation is blueshifted, but instead it can be redshifted even if it is approaching, provided it is moving fast enough and has some transverse component. Hard to believe but true!
4. If that isn't strange enough, there is yet another mechanism in play–the "gravitational redshift." The general theory of relativity says that, near a mass, there is a gravitational well. When light passes near a large mass its wavelength is altered because of time dilation. The gravitational field changes time itself, and shifts the wavelength of the light passing through it.

Astronomers find redshift extremely useful. A few examples are measurement of the rotation of galaxies, rotation rates of planets, velocities of interstellar clouds, and movements of the Sun's photosphere. The redshift of microwave background radiation, with its tremendous z of 1000 (the shifted wavelength is 1000 times the original wavelength), shows the state of the universe 13 billion years ago. The redshift of distant galaxies shows the expansion history of the universe, and that the expansion is now accelerating.

"Redshift" spectroscopy is one of the most productive techniques in an astronomer's tool kit.

---

## Gravitational Waves

Joseph Taylor and Russell Hulse, astronomers at Princeton University, have been making measurements of an extraordinary binary star since 1982. It consists of two incredibly dense neutron stars rotating around each other very rapidly. The binary pair, each of whose stars has a mass about 1.5 that of our Sun, is called PSR 1913+16.

What they have been measuring is radio pulses from one of the two stars, a pulsar. Their measurements provide evidence of "gravitational waves." Gravity effects are due to curvature of space (which is better called space-time), and ripples in space-time are gravitational waves or "gravitational radiation." These waves, which travel at the speed of light, are caused by acceleration of masses, as explained below.

A pulsar is a very magnetic rotating neutron star that emits radio signals in short, regular bursts. This one puts out about 16 pulses per second. Taylor and Hulse have measured the frequency with the amazing accuracy of one part in 10 to the 14th power! In 1993 they received a Nobel Prize in physics for their work.

The two stars complete a rotation around each other about three times per Earth day. There is Doppler shift in the frequency of the radiation; the frequency increases as the star approaches the Earth and decreases as it moves away.

There are three frequencies involved: (a) the "carrier" frequency of electromagnetic radiation, (b) the pulse repetition rate (chirps) of about 16 pulses per second, and (c) the frequency of rotation of the two stars around each other, which has a period of about eight hours.

The frequency of rotation of the binary pair has been increasing very slightly, which indicates that the stars are getting closer together, and which also means that energy is being lost from the system.

The idea of gravitational waves comes from Einstein's general theory of relativity–the one that deals with acceleration. The measured rate of increase in frequency of rotation is within one percent of what would be predicted by Einstein's theory, so the energy is probably being lost by radiation of gravitational waves.

That change of frequency is accompanied by "perihelion precession," which is a creeping of the orbit's major axis of rotation. Precession results in a complicated way from gravitational pulls. The orbit of this binary pair precesses 30,000 times faster than the precession of the perihelion of Mercury, which gave earlier confirmation of the curvature of space-time, so this newer confirmation of space-time curvature is much more convincing.

In 2005, observations of another binary pulsar, PSR J0737-3039, were reported that show that the system's orbit shrinks seven millimeters per day. This provides further confirmation of existence of gravitational waves.

There is an analogy here with electromagnetic radiation. We know (from Maxwell's equations) that electromagnetic radiation is produced by acceleration of electric charges. For example, in the familiar Bohr model of the atom, when an electron falls from one energy level to a lower one, a pulse of electromagnetic radiation comes out.

In a similar way, acceleration of masses produces gravitational waves. Rotation of binary stars around each other involves continuous change of direction, which, of course is a form of acceleration. That acceleration of the massive stars produces the gravitational waves in space-time.

Gravitational waves here on Earth are very weak and have never been directly observed. Astronomers are attempting three or four other types of astronomical experiments to detect the existence of gravitational waves, without much success so far. When more detectors of gravitational waves are devised, we might have not only astronomy of light and other electromagnetic waves, but also astronomy of gravitational waves.

References:

For "The Mystery of DarkEnergy"

1. The Facts On File Dictionary of Astronomy, by John Daintith and William Gould, Infobase Publishing, New York. Hardcover, $60.00.
2. http://en. Wikipedia.org/wiki/dark_energy
3. The Road to Reality, by Roger Penrose. Page 772. Vintage Books, New York.

For "Dark Energy" & "More about Dark Matter"

4. *Zwicky, F.* "On the Masses of Nebulae and of Clusters of Nebulae." *Astrophysical Journal 86: 217, 1937.*
5. *Zwicky, F..* "Die Rotverschiebung von Extragalaktischen Nebeln." *Helvetica Physica Acta 6: 110–127, 1933*

6. "Some Theories Win, Some Lose." NASA, using the WMAP.
7. "Dark Matter." Wikipedia, the Free Encyclopedia.

For "Redshift and What It Tells Us"

8. http://www.sopof.gsfc.nasa.gov/stargaze/Sun4/adop1htm
9. http://www.electric-cosmos.org/arp.htm
10. http://www.ucla.edu/~wright/doppler.htm
11. http://en.wikipedia.org/wiki/redshift
12. http://en.wikipedia.org/wiki/dielectricconstant
13. http://www.arachnoid.com/sky/redshift.html
14. http://panisse.lbl.gov
15. http://astSun.astr.virginia.edu/-jh8h/glossary/redshift.htm

# Chapter 13
# Light

## The Incredible Speed of Light

Speed Light zips ahead at a mind-boggling 186,000 miles per second in a vacuum. It's almost as fast in air, and about two-thirds that fast in glass.

Lenses Because of the difference in speeds, light is refracted if it strikes a glass-to-air interface at any angle other than straight on. A lens works because the interface is curved. The part of the wave front that arrives at the glass first, starts traveling at the lower speed earlier, so the light changes direction (picture a big family walking abreast and marching diagonally into a very muddy field.) The refractive index of a medium is inverse to the speed of light in it.

Faster? Is anything faster than light? Einstein said that as a body is accelerated its miss increases, and that it would increase toward infinite mass if its speed could approach that of light. That means that the force that would be required to accelerate a mass to the speed of light would be infinite; it's unlikely that any mass goes that fast. Nevertheless, one Case WRU cosmologist points out: "We know that many things in the universe are receding, and if something were to recede faster than the speed of light we would never be able to see it, and wouldn't know anything about it." I wouldn't bet the farm on anything's being that fast.

Relativity The speed of light is independent of the velocity of the observer, so there can't be any relative motion between an observer and a light beam. That's tantalizingly hard to picture. It is the basis of the theory of relativity.

Duality As most people know, light has a schizophrenic nature, sometimes acting like waves and sometimes like particles. Reflection, refraction, interference, and polarization of light can be explained best in terms of wave motion. But when light interacts with matter, involving absorption and emission, it behaves more like particles.

Fields Light waves need both electric and magnetic fields to travel, just as surely as a person needs both diet and exercise to lose weight.

Gravitational Lenses Gravitational lensing of light is affected by the curvature of space near a mass. A gravitational field bends light, so a smoothly distributed concentration of mass such as a galaxy or cluster of galaxies focuses light rays as a glass lens does. Einstein and Lodge wrote about that idea theoretically in 1919,but the first known gravitational lens was not observed astronomically until 1979.

---

## A Long Ride on a Light Wave

Imagine we're sitting on a magic carpet and riding toward the Earth on a light wave from Sirius, our brightest star. To make it easier to imagine, let's change the wavelength from a visible 0.00006 centimeters to ten feet. The electromagnetic radiation is then a short radio wave, but the behavior is the same.

If the wave is horizontally polarized, the electric field is horizontal, and alternating left and right in sine wave fashion.

The magnetic field is vertical, and alternates up and down. We travel straight ahead, with the flow of energy, at a speed of 186,000 miles per second.

If we come to a cloud of cosmic dust, we might enter the cloud at a slant. If so, our direction turns very slightly into the cloud, because of refraction at the interface between empty space and the cloud. Our wave slows down ever so slightly when we enter the cloud, because our speed is inversely proportional to the square root of both the dielectric constant and the magnetic permeability of the medium.

Our wave spreads out "conically," filling a cone, as it travels along, so it keeps getting weaker. Its power density diminishes according to the inverse square law. However, this doesn't mean that the electric field diminishes in inverse proportion to the square of the distance traveled, because the power density of the wave is the electric field strength multiplied by the magnetic field strength. The electric field strength itself diminishes inversely as only the first power of the distance, and so does the magnetic field. It is their product that diminishes according the inverse square law.

Of course, we can't really take this ride, because to get up to the speed of light we'd have to be accelerated, and we'd put on weight. The closer we got to the speed of light, the greater force it would take to speed us up by still another mile per hour. We'd approach infinite mass as we approached the speed of light so the force that's trying to accelerate us would have to approach infinity.

Are we there yet? After 8.7 years we arrive at Earth on our wave of light from Sirius. And a lot of astronomers on this planet will notice us as being starlight, after this amazing trip.

# Light–Is It a Wave or Particles?

George (astronomy student): Light is a fascinating mystery to me. I see it every day, but I don't really know what it is.

Professor, (Marie): 150 years ago scientists were sure that light was a wave of energy. And it is, for some purposes but not all.

George: Then, what is light?

Professor: There is a lot of evidence that light consists of waves, and plenty of other evidence that it is particles. Sometimes it is referred to as a wavicle.

George: When we say light consists of particles, do we mean it has mass?

Professor: No. We mean the light's energy is in little bundles, that is, photons or quanta.

George: How can light consist of waves and particles at the same time without converting back and forth?

Professor: There are different ways that light reveals itself, depending on what observations we make. When we observe an interference pattern it reveals its wave aspect, and when we observe its photoelectric effect it reveals its particle aspect.

George: What experimental evidence is there that light is waves?

Professor: The evidence that light and other electromagnetic radiation are waves are these phenomena:

Interference patterns.

Diffraction patterns

Polarization.

George: I've seen all of those wave phenomena, and they are persuasive. What experimental evidence is there that light is particles?

Professor: Several phenomena give experimental evidence that light and other electromagnetic radiation are particles:

- Black body radiation. A black body emits light when it is heated. Max Planck had to concede that light was emitted in packets in order to explain the observed shape of the distribution curve of emitted wavelengths.
- The photoelectric effect. Albert Einstein used Planck's idea of photons to explain the photoelectric effect. When the light intensity on a photoelectric surface is increased, the energy of the emitted electron doesn't change; but when the intensity is doubled, the number of electrons that are emitted is doubled. Einstein received a Nobel Prize for this.
- The Compton effect. At X-ray frequencies and above, when a wave strikes an atom and an electron is released, the frequency of the remaining scattered wave is decreased. The incoming X-ray must have been a photon. Compton also received a Nobel Prize.
- Pair production. In 1932 Carl Anderson found that if a gamma ray has enough energy (at least 2 $mc^2$, where m is the mass of an electron), when it strikes a nucleus, it can produce an electron and a positron, each having a mass corresponding to $mc^2$. The gamma ray must have been in particle form. Anderson received a Nobel Prize for this work.

George: I'm losing contact with the real world, professor! I find it hard to imagine that light is waves and particles at the same time.

Professor: The idea is different from our usual way of thinking because we get our experience from the macro world. But the micro_world is different, and is hard to accept intuitively.

George: Are we going to get into this next semester?

Professor: Yes. When we study quantum mechanics you'll be convinced that light is both waves and particles at the same time. Until then, try to put your macro_world experience aside, George. Get over it!

---

## Light's Electric and Magnetic Fields

High School Student: When you cut an apple the pieces stay where they land, but when you generate light it travels. Why?

Physics Teacher: Well, that's not so strange–water waves travel too. The water doesn't travel much; it just goes up and down, but the wave energy travels. Sound waves travel too–in air, for example.

Student: But air and water waves aren't like light. Air and water have mass and restoring forces, but light can travel in a vacuum.

Teacher: Yes. It's because a vacuum, like other media, can store energy in magnetic and electric fields. That energy can go back and forth between the two fields.

Student: But how do those fields create light?

Teacher: When a source, say a star, emits light energy into space, the energy is in electric and magnetic fields, and it travels. To illustrate what happens, your left arm can represent an electric field and your right arm can represent a magnetic field. Hold both arms in front of your chest, with the left arm vertical and the right arm horizontal, so they cross near the wrists.

Student: OK.

Teacher: When light is generated, an electric field goes positive (in an upward direction) in the left arm, then a half period later it goes negative, in sine wave fashion. The electric field doesn't move in the arm; the field just grows stronger and weaker and reverses.

In the same way the magnetic field pulsates in the right arm, first growing stronger with a leftward polarity, then subsiding and growing stronger with a rightward polarity.

The electric and magnetic fields are perpendicular to each other. In this example the direction of propagation can be either in the direction you're looking or backwards. The fields are called transverse fields because they are perpendicular to the direction of light propagation. The wave equations are too long a story for now, so trust me that these fields do move.

If the electric field is vertical the wave is said to be vertically polarized, but the polarization may rotate as the wave travels.

Student: Why does the light get weaker as it travels?

Teacher: Its energy spreads out. If a certain amount of energy in a first area spreads out to a larger area the energy density farther away is reduced by the ratio of areas. Twice as far from the source, the area is four times as great. The energy density is inversely proportional to the square of the distance from the source. That's the inverse square "law" of light attenuation.

Student: Good! How fast does light travel?

Teacher: The speed of light is 186,000 miles per second in a vacuum but slower in other media, for example in air or water. The velocity depends on the electric "permittivity" and magnetic "permeability" of the medium.

Student: Why does light bend when it enters water diagonally?

Teacher: That's called refraction. Light travels with different velocities in different media, and when light passes into water it slows down. If a flashlight beam enters the water on a slant, the lower part of the light, which strikes the interface first, is slowed down earlier than the upper part of the light, which arrives at the interface later. Because of that, when the wave reforms itself in the water, the direction of travel is bent so as to be closer to a perpendicular to the interface.

Student: Is refraction unique to light?

Teacher: No. The equations of electromagnetism, such as Maxwell's equations, apply to all frequencies of electromagnetic radiation. That includes radio, TV, infra-red, visible light, radiant heat, X-rays, gamma rays, cosmic rays, etc.

Student: OK. Now, light goes in a straight line doesn't it?

Teacher: Usually, but it curves when it passes near a mass. Albert Einstein showed that space itself is curved near a mass, and light follows the curvature of the space.

Student: Thanks for taking the time to explain all that, Mr. Lehrer.

Teacher: You're welcome, and I'm giving you an A for asking those great questions.

---

## How Light Travels

High School Student, Britney: Good news! I've been accepted as an astronomy student at Ohio State next year. You told me about electromagnetic fields last month, but you didn't have time to tell why light travels instead of just staying where it is generated.

Physic Teacher, Mr; Lehrer. Congratulations, Britney. There are a wave theory and a corpuscular theory of light, and are both valid.

Student I'm wondering why light travels at all, Mr. Lehrer.

Teacher: You're asking "why." A purist would say that an important difference between science and other fields is that science would ask "how does light travel?" To be meticulous, a scientific theory telling "how" has to be testable, so that it can be proven false if it is false. There are some fields, such as astrology, in which the "how" can't be tested, and its followers accept it on belief.

Student OooKaaay, then, how does light travel?

Teacher: It is easy to visualize with the corpuscular theory, in which it travels as little bundles of energy, because we've all seen a baseball game. It's harder if you think of light as a wave.

Student I understand the baseball idea, so let's concentrate on waves.

Teacher: Well, we use what is called a Poynting vector, which shows the direction and strength of travel of the flux of wave energy. You know about a vector–it points the way that something is going and its length shows the magnitude.

Student So the energy flux of an electromagnetic field can be represented by this Poynting vector?

Teacher: Exactly. A Poynting vector is perpendicular to both the electric field and the magnetic field of the wave, and it points in the direction the energy travels. Its length is proportional to the electric field strength multiplied by the magnetic field strength.

Student Then light energy moves in the form of pulsating electric and magnetic fields according to a Poynting vector. How fast does the energy travel?

Teacher: It goes at an amazing 300,000 kilometers per second. And incidentally, the speed of light is the same for all observers, irrespective of the observer's own speed! Incredible but true.

Student What do you mean about the speed of light being the same regardless of the speed of the observer?

Teacher: Here's an example. Suppose a person on a moving railroad car measures the speed of light from his flashlight. At the same time, an observer on the ground somehow measures the speed of the same light. They measure the speed as being the same even though the flashlight is moving. It works because, according to Albert Einstein, time is running at a slightly different rate for the person on the rail car than for the person on the ground.

That doesn't mean there is no Doppler-like effect on the frequency of a signal. For example, the light from receding stars is red-shifted, as you know.

Student: Better stop now Mr. Lehrer–you're boggling my mind. Thank you.

Teacher: See you in class, Britney.

# Chapter 14
# Einstein, Michelson and Morley

## Einstein's Amazing Creativity

Albert Einstein contributed more to science than any other person. His theories are hard to visualize because they are so counter-intuitive.

Today, despite the fact that the transmission of television signals is hard to imagine, we accept the idea readily because we can see the TV picture. But such Einstein concepts as "time is different depending upon your speed," and "starlight bends slightly as it passes by the Sun" are astonishing and never witnessed in our daily routines. In fact, it has been very hard to test his theories because they are so difficult to measure here on Earth.

Born at Ulm, Wurttemberg, Germany in 1879, Albert Einstein was taken by his family to Munich when he was six weeks old. He went to school in Zurich Switzerland and received a college diploma to teach physics and mathematics, which he didn't do until he became a professor much later.

Einstein renounced his German citizenship and in 1901, became a citizen of Switzerland. In 1902 he got a job as a technical assistant examiner at the Swiss patent office.

In 1908 Einstein met and married his first wife Mileva Maric, a mathematician. He once said she was his "equal." Some feminists said his wife was the thinker and he only the scrivener. It's possible that she did play a part in one of his preliminary papers on relative motion. They had three children, the first of which was born before their marriage and put up for adoption.

Einstein based his special relativity theory (1905) on the concepts that (a) the speed of light is the same for all observers and (b) the laws of nature are the same for all observers who move with a constant speed relative to each other.

Thus his special theory of relativity deals with situations in which masses have constant relative speeds. He averred that the speed of light is fixed and not relative to the speed of the observer. This was a departure from Newton's laws of motion.

One of his 1905 articles led to his famous equation $E=mc^2$. It says that the energy (E) of a body at rest equals its mass (m) times the square of the speed of light ($c^2$). The shocking meaning of this equation is that a body has energy even at rest, apart from its potential and kinetic energies. The at-rest energy of a body exists because of its mass. Also, it says that any increase in the energy of a body must result in an increase in its mass, so high velocity means high mass also.

The amount of energy represented by even a small mass is tremendous (as shown by the atomic bomb), compared with the energy released by chemical explosions.

His general theory of relativity (1916) involves masses that are accelerating relative to each other. One surprising consequence of the general theory is that uniform acceleration and a gravitational field are equivalent. His theories were ridiculed at the time of their publication. However, both of the relativity theories have survived all of the experiments devised to test them so far.

In 1914 Einstein moved to Berlin to accept a position, and became a German citizen.

He divorced Mileva in 1919 and married Elsa Einstein Lowenthal in the same year. She was his first cousin on his mother's side and a second cousin on his father's side. Elsa nursed him when he had a partial nervous breakdown.

Although the most famous of his accomplishments are his special and general theories of relativity, he did a lot more. Other work was in the fields of Brownian movement, Bose-Einstein statistics, photoelectric effect, unified field theory and many others.

Einstein was awarded a Nobel Prize for his outstanding contributions to physics generally, but the Nobel Prize specifically named his work on the photoelectric effect (the photon theory). His other contributions, which were much greater, were not listed in the prize because those theories were still controversial.

Einstein received many honorary doctor's degrees. He was also offered the presidency of Israel, but he declined, saying that he was a physicist, not a politician.

His theories were rejected by some critics because he was Jewish. One group issued a paper called "One Hundred Authors Against Einstein." He retorted "If I were wrong, one would be enough."

He renounced German citizenship in 1933 because of politics and moved to Princeton New Jersey to work at Princeton University. Elsa died there in 1936. In 1940 Einstein became a United States citizen.

He retired from Princeton in 1945 and died in 1955. His efforts in the last two decades of his life to devise a unified field theory were helpful but largely unsuccessful. Today physicists are working on "string theory," which is a possible fulfillment of his goal to find a unified theory.

At the end of the 20th century Einstein was named Man of the Century by Time magazine.

He died in 1955. In his will he said he didn't want a funeral, a grave or a marker. Perhaps he thought they would be desecrated. His body was cremated and the ashes cast into a nearby river.

Forty-four years after his death a poll of 100 leading physicists named Albert Einstein "the greatest physicist ever." A mantle of "great philosopher" will never descend upon his shoulders because his philosophical ideas, whether correct or not, were not new with him.

The pathologist who performed an autopsy on Einstein preserved his brain. In 1999 the brain was studied by others who said that one region was missing, and its absence was compensated by another region that was 15% wider than average and was related to mathematics, spatial visualization and movement.

Albert Einstein is enshrined in the world's memory for his brilliant contributions to knowledge of our universe.

---

## Albert Einstein's Philosophy

Albert Einstein's philosophy is often quoted. His own words describe it.

No Personal God

In 1954 he wrote in a letter "I do not believe in a personal God and I have never denied this but have expressed it clearly." He said "I do not believe in immortality of the individual, and I consider ethics to be an exclusively human concern with no superhuman authority behind it."

"The most important human endeavor is the striving for morality in our actions." He also said "Only morality in our actions can give beauty and dignity to life."

He once told a Rabbi "I believe in Spinoza's God, who reveals Himself in the lawful harmony of the world, not in a God Who concerns Himself with the fate and the doings of mankind."

God and Creation

In the pronouncements above Einstein sounded like an atheist. Nonetheless, in other statements he sounded religious, but did not quite contradict the statements above.

"I want to know how God created this world. I am not interested in this or that phenomenon, nor in the spectrum of this or that element. I want to know His thoughts; the rest are details." It was characteristic of Einstein to want a unified overview.

"Science without religion is lame, religion without science is blind."

Einstein was skeptical about the probabilistic aspects of quantum theory because he believed in deterministic physics—causes and effects. He wrote "I, at any rate, am convinced that He (God) does not throw dice." Niels Bohr replied with "Stop telling God what He must do."

"God is subtle, but He is not malicious."

In a private letter he made a harsh and very controversial statement: "The word God is for me nothing more than the expression and product of human weakness, the Bible a collection of honorable, but still purely primitive, legends which are nevertheless pretty childish." That seems insensitive and politically incorrect. He was castigated and even threatened with death for a few of his views that were made public.

Uncertainty

In an interview with Time magazine, Einstein said, "I am not an atheist and I don't think I can call myself a pantheist. We are in the position of a little child entering a huge library filled with books in many different languages. The child knows someone must have written those books. It does not know how. The child dimly suspects a mysterious order in the arrangement of the books but does not know what it is. That, it seems to me, is the attitude of even the most intelligent human being toward God." Einstein had the onerous bugaboo that he "does not know" an answer to the God mystery. Many people may have that problem and not like the uncertainty.

Summary

Giving Einstein full faith and credit that he said only what he truly believed, all of his statements must be taken together. His integrated philosophy sounds like deism, which strongly advocates morality but disavows any unnatural control over human affairs. He said "I believe in Spinoza's God" Spinoza believed God exists only philosophically and that God is abstract and impersonal.

Einstein summarized his own ideals in 1954: "Kindness, Beauty and Truth."

Later Life

Being a pacifist, Einstein was appalled by the use of the atomic bomb in World War II, although he had contributed to the development of nuclear power. He collaborated with Bertrand Russell and Albert Schweitzer to lobby nations for stopping nuclear testing and future bombs.

Some recent publicity alleges Einstein was a womanizer.

He died in 1955. In his will he said he didn't want a funeral, a grave or a marker. Perhaps he thought they would be desecrated. His body was cremated and the ashes cast into a nearby river.

Forty-four years after his death a poll of 100 leading physicists named Albert Einstein "the greatest physicist ever." A mantle of "great philosopher" will never descend upon his shoulders because his philosophical ideas, whether correct or not, were not new with him.

---

## Gravitational Lensing

Evidence of Gravitational Lensing

The presence of mass warps the space around it and the path of light that passes near it. If the light from a distant object such as a quasar passes near a large mass or a distributed concentration of masses such as a galaxy or cluster of galaxies, the galaxy can sometimes act as a lens and focus the light rays. Focusing by gravitational "lenses" is usually crude and distorted, as you would expect, but it is nevertheless a useful effect.

Usually a galaxy that causes lensing is not visible itself. There are various types of gravitational lensing. When a galaxy or cluster of galaxies causes lensing it is called macrolensing, and when a single object crosses the light path and causes lensing it is called microlensing. When the distant galaxy in the background is also extended, then the lensed images are smeared out into long luminous arcs several arc seconds long. If the extended background source is exactly aligned with a symmetrical lens, the lensed image takes the form of an "Einstein ring." Multiple images of the remote object can be formed.

Invisible dark matter can be a gravitational lens, and the presence and density of such dark matter can be estimated by its lensing effects. The first known gravitational lens was of a double quasar (two spaced-apart images of the same quasar). By1993 only seven possible double quasars had been found, and one triply imaged quasar. Four quadruple lensing quasars, including the "Cloverleaf" and the "Einstein Cross" have been observed.

If the light paths are of unequal length and the quasar itself is variable, light from one image of a double quasar arrives at Earth at a different time than light from the other image of the same quasar. A difference in the time at which a brightening is seen in two images may be measured, and can be several days or even years.

Quasars, which are favorite objects for gravitational lensing, were discovered in 1963; several thousand quasars are now known. They have very great redshifts. They are extremely luminous; many have absolute magnitudes brighter than minus 27.

Theory

Gravitational lensing of light is caused by the curvature of spacetime near a mass. Einstein and Lodge wrote about the theory of gravitational lensing in 1919. The eccentric Swiss-American astronomer Fritz Zwicky tried to apply the idea to astronomy in1937. The first known valid gravitational lens effect was not actually observed astronomically until 1979. The general theory of relativity involves the equivalence of gravity and acceleration. Curvature of spacetime is the explanation for that equivalence. Light takes the shortest path within the curved spacetime geometry, just as a great circle route is the shortest path of air travel from Cleveland to London irrespective of lines of latitude and longitude.

Information Obtained from Gravitational Lensing

There is further confirmation of the theory of general relativity because of the deflection of light around masses. Sometimes verification that a quasar being focused upon is much more distant than the galaxy focusing it, shows that the quasar's great redshift is "cosmological." (It isn't "gravitational redshift" [Einstein shift], which is redshift due to loss of energy incurred by light when the light escapes from a large mass.)

Gas around the focusing galaxy can create the distant quasar's absorption-line spectrum, revealing a lot of information about the gas around the focusing galaxy.

Some images of remote objects are temporarily made much brighter by magnification as a gravitational lens passes in front of them, making some previously invisible stars visible.

When dark matter acts as a gravitational lens it helps astronomers to estimate the density of dark matter in the universe, which is very important.

Hubble's constant can be computed by the differences in travel time of two images of the same remote object, e.g., by the time delay between appearance of an event as it reaches the observer.

Events that are seen in one image of an object can be anticipated and studied in its companion images after the known time delay. For example, an explosion seen in the earlier image will appear later in the other image, so astronomers can get ready to study the event.

Gravitational lensing has already become a very useful and important tool for astronomers, and it is evolving rapidly.

# Michelson and Morley, Ethical Giants

Michelson and Morley were the first to successfully compare the speed of light in the direction of the observer's travel with the speed at right angles to the travel to prove or disprove the existence of an "ether." They found no difference in the speed of light between the forward and transverse directions.

Ever since Aristotle, the scientific community had believed that electromagnetic waves traveled in a stationary ether, so for an observer on the Sun-orbiting Earth, the ether would appear to be like a river current. For a light that was launched on Earth, the ether current would decrease the velocity of light in the forward (upstream) direction (relative to outer space) and increase it in a backward direction, with the decrease being unequal to the increase. There would be a net increase in total <u>travel</u> time.

The transverse light beam, on the other hand, would hardly be affected by the current, and it would travel over and back in a shorter time than the upstream/downstream beam. A measured difference in velocities in these 90° - apart directions would have proven that an ether exists. But this experiment said NO!

It's the psychology of the situation that's amazing. Wouldn't you think they would have decided they did something wrong, and slink off? But to their professional credit, Albert Michelson and Edward Morley courageously reported this "negative" experiment.

Michelson and Morley were professors at Case and Western Reserve, respectively. Michelson received a Nobel Prize in 1907 for the work he did before, during and after this famous experiment.

The test was done in a Cleveland basement with apparatus floating on a granite slab that was rotatable in a large basin of mercury. Because the Earth travels around the Sun, that basement was sailing along through the "ether" at 584 million miles per year, or 65,000 miles per hour.

You'd have to see a diagram to understand the apparatus, but here is a very short description. A monochromatic light source sent a light beam forward to a mirror, where it bounced back and was reflected sideways to a detector. From the same light source, a beam splitter provided a second beam that went transversely, to a mirror 90° away from the first, and that second beam came back to the same detector.

The detector was an interferometer, and any difference in the two trip times would show up as interference fringes because the beams would then have slightly different wavelengths. The speed of light is 10,000 times greater than the observers' orbital speed, but because light has such a short wavelength, a light interferometer can detect a very small change. But the fringes showed that the trip times were identical.

Albert Einstein may or may not have recognized in 2006 that Michelson and Morley's 1887 experiment would fit his special theory of relativity--that the speed of light is the same <u>irrespective</u> of the speed of the observer. The reason for this counter-intuitive concept is that time marches at a different rate depending on the speed of the observer--at a rate to make the speed of light look the same regardless of the observer's speed.

Hendrik Lorentz tried to explain the test results by saying that the length of the apparatus was "contracted" by the ether. Although the reason he gave isn't correct, the Lorentz contraction equation is true for other reasons, given in the later special theory of relativity.

Michelson and Morley reported test results that were contrary to what the entire world believed. Their scientific integrity has made them role models.

---

## The Michelson-Morley Experiment

This is the second part of four sections. Albert Michelson and Edward Morley (M & M) conducted an experiment to test the "ether hypothesis," which was that light and other electromagnetic waves are propagated through some kind of transparent fluid that fills all of space.

This fluid was supposed to act as a medium for the propagation of light analogous to the manner in which sound is propagated through matter. If the then-popular "ether' theory were true, motion relative to the ether should be detectable. Michelson designed a special interferometer apparatus that would detect the effect if it existed.

The M & M experiment of 1887 was the first optical experiment performed with sufficient accuracy to be able to detect motion of the Earth through the ether (if there were an ether). M & M showed that no ether effect was detectable, so the theory was probably incorrect.

The Earth travels around the Sun, therefore through the "ether," very fast. (Even speeds due to the Earth's rotation on its axis are insignificant compared with the speed of the Earth in its orbit.) The ether theory said that they would find a difference in the velocities of light in different directions, if one was in the direction of the Earth's motion in its orbit (i.e., in the direction of the Earth's travel through the ether) and the other was at a right angle to that direction. Instead they found that the velocity of light was identical irrespective of its direction.

If the ether propagation theory were correct, differences in round-trip time of reflected light would occur, because when a light source is moving in a fixed medium the change of the light's trip time when going outbound is not exactly equal and opposite to the change of return-trip time after reflection.

On the other hand, the outbound- and return-trip times are equal and unaffected if the light goes at right angles to the motion of the source. Consequently, if the ether were a fixed medium, the time of the passage of the light should be greater for a to-and-fro path in the direction of the Earth's motion than for a to-and-fro path, of equal length, at right angles to the direction of motion.

Here is an analogy to show that the motion-induced changes in the outgoing and returning travel times do not exactly offset each other. Say a boat goes 30 mph in a 2 mph river. The boat's downstream velocity is 32 mph, so the time required to go 1 mile downstream is 1/32 hour, which is 0.03125 hour. The boat's upstream velocity is 28 mph, so the time to go 1 mile upstream is1/28 hour, which is 0.03571 hour. A 1-mile round trip takes 0.03125 + 0.03571 = 0.06696 hour.

If the river weren't flowing at all, a round trip would take only 2/30 hour, which is 0.06667 hour. Note that this is less than 0.06696 hour. This shows that a round trip takes longer on a flowing river than on a stationary one. In this analogy the boat is like the light beam and the river is like the motion of the Earth along its ecliptic.

This principle was used by M & M to test for the existence of an ether, as will be discussed in the next article. The next article will describe the apparatus M & Mused to conduct this famous experiment. The third installment on M & M will tell more about the people involved and Einstein's solution of this apparent dilemma about the speed of light.

## More about the Michelson-Morley Experiment

This is the third of four parts. Albert Michelson and Edward Morley (M & M) conducted an experiment to test the popular "ether hypothesis," which was that light and other electromagnetic waves are propagated through some kind of transparent fluid that fills all of space.

Michelson was a Case Tech professor and Morley a Western Reserve University professor. This tells how they did their famous experiment, which was done in a basement in Cleveland. It is difficult to describe the apparatus without diagrams, but here goes.

Schematically, Michelson's interferometer consists of two arms set at right angles to each other in an L shape. Each arm has a mirror at the far end. At the intersection where the arms are joined there is a half-silvered mirror,-toward which a light beam is directed. The half-silvered mirror splits the light beAM in two. Each half of the split beam travels down one of the arms and is reflected back by the mirror at the end of its arm.

Thus, a monochromatic beam of light is divided by a half-silvered mirror, sent along two orthogonal paths, and later brought back together by further mirrors. When the two beams are recombined, they interfere in such a way as to produce a pattern of fringes. The pattern produced depends upon the difference in elapsed time required for the two beams to make the round trip.

If the entire apparatus is rotated through 90 degrees, the roles of parallel and perpendicular arms are reversed; the fringe pattern would shift if the difference in travel times changed. As explained in the previous article, if they are moving in a medium such as ether, the beam traveling crosswise to the direction of flow will take less time to make a round trip.

The interferometer was on a heavy block of stone mounted on a disc of wood. The disk floated in a tank of mercury, so it could be rotated smoothly, arms and all. To increase the light path each of the interfering beams was reflected back and forth several times; four mirrors were used to replace each of the single mirrors in the simple form of interferometer, to provide a total path length of 1100 cm.

The apparatus was first set so that one path lay in the direction of the Earth's orbital motion, then the apparatus was turned through a right angle, so that the beams of light interchanged their roles. The beam which, in the parallel first position, traveled the slower, would travel the faster in the transverse second position. The two beams in the interferometer formed a system of interference fringes, and this fringe system could shift when the apparatus was rotated so as to pass through the two positions.

According to the ether hypothesis the interference fringes formed in this interferometer should have moved as the apparatus was slowly rotated a quarter turn. They didn't. The expected fringe shift due to the Earth's orbital speed was four-tenths of a wavelength; however, no shift as large as even four-hundredths of a wavelength was observed.

Moreover, the experiment was carried out both at midday, when the Earth's velocity relative to the Sun was in the plane of the apparatus, and at 6 pm, when the velocity was normal to the plane of the apparatus.

This marvelous experiment posed an apparent dilemma about the speed of light. The next article tells about Albert Einstein's ultimate solution of the puzzle, and more about the people involved in the work. References:will be given at the end of the next article.

## Still More on the Michelson-Morley Experiment

This is the last of four parts. Michelson and Morley (M & M) conducted an experiment to test the popular ether hypothesis, that light is propagated through some kind of transparent fluid. Their interferometer apparatus was first set so that one path lay in the direction of the Earth's orbital motion, and the other at a right angle to the orbital motion. A pattern of interference fringes was formed when the two beams were recombined in the Michelson interferometer. Then the apparatus was turned through 90 degrees, so the beams of light interchanged their roles. If one beam had traveled slower when in the (parallel) first position, it would have appeared to travel faster when in the (transverse) second position.

The experiment depended upon the fact that a round trip in the direction of motion relative to a medium (ether) would take longer than a round trip transverse to the direction of motion. It is hard to see why. The reason is, for the round trip in the direction of motion, the travel time upstream is of course greater than the travel time downstream; i.e., the moving object naturally spends more time on the upstream portion of the trip than on the downstream portion.

But the upstream travel time all occurs at a lower speed than does the downstream travel time, so the average speed for the whole round trip is slightly greater than for a transverse round trip. The time gained while going downstream isn't enough to completely cancel the loss while going upstream. Incidentally, this can easily be proven by algebra.

If the ether theory were correct, the fringe pattern would have shifted when the apparatus was slowly rotated so as to pass through the two positions. However, the velocity of light was found to be the same in the two directions.

This experiment raised an apparent dilemma about the speed of light. Upon learning that the velocity was the same in all directions, other scientists advanced various theories to explain the results.

Many further experiments were conducted to investigate possible explanations. None of the ancillary experiments ultimately contradicted the M & M result. lt now seems fairly certain that no motion of the Earth through something that might be called an "ether" can be detected by either light or radio waves

Einstein put forth his special theory of relativity in 1906, and it explained the phenomenon. lt says that the speed of light is always the same, regardless of the motion of the observer, and therefore is the same in each direction along each arm of the interferometer. The velocity of light is always the same, as measured by any observer (of a set of observers) who is moving with constant relative velocity.

Einstein's theory of relativity provides the only fully consistent explanation of the M & M results. In another experiment using a different kind of interferometer mounted on a telescope, Michelson found the angular diameter of Betelgeuse to be 240 million miles. It was a measurement for which no telescope alone had nearly enough resolution. Michelson also measured the speed of light very accurately.

In 1907 Michelson received a Nobel Prize in physics for his accomplishments.

Michelson and Morley were great scientists. They reported their startling measurements and conclusions just as they found them, and they were right.

References:
For "Einstein's Amazing Creativity"

1. www.spaceandmotion.com_einstein_biography
2. http://nobelprize.org/nobel_prizes/physics
3. http://scienceworld.wolfram.com/biograhy/einstein.html
4. http://en.wikipedia.or/wiki/Theory_of_relativity For "Albert Einstein's Philosophy"
5. refspace.com/quotes/s%253AO "Popular Quotes." Retrieved 2009
6. http://en.wikipedia.org/wiki/Albert_Einstein
7. http://home.pacbell.net/kidwell5/aebio.html
8. http://scienceworld.wolfram.com/biography/Einstein.html

# Chapter 15.
# Technicalities of Telescopes

Dick and Jane, Amateur Astronomers

Hi, my name's Dick.

I'm Jane.

Dick: I think you live across the street from my house.

Jane: Yes.

Dick: I liked tonight's talk at the astronomy club. I learned a lot about photographing the images that are in a telescope. How is an image created in the first place anyway? Do you know much about it?

Jane: A little. Well, as an example, let's say that the star Pollux is 30 degrees above the horizon and Castor is just above it at 33 degrees, and both of them are within the telescope's field of view.

Dick: OK.

Jane: Both stars are hundreds of millions of miles away and the telescope is only one ten-thousandth of a mile in diameter. So all of the rays from Pollux that enter the aperture are essentially parallel to each other, even though some enter near the top of the aperture, others near the bottom, some at the left, some at the right, and so on.

Dick: Is electromagnetic radiation really rays?

Jane: No, it isn't, but rays are easier to talk about than wave fronts or photons.

Dick: What makes the image then?

Jane: To form an image, all of the captured rays from Pollux, regardless of what part of the aperture they enter, are focused to come to one point at one image plane. That trick may seem like magic; the shape of the primary lens (or mirror) gets the credit for it. The image plane is at right angles to the optical axis, and for these very remote targets it is located a focal length away from the primary lens, and not far from the eyepiece.

Dick: Doesn't Castor also have parallel rays at the telescope?

Jane: Yes. All of the rays from Castor that enter the aperture are also parallel to each other, but they are far from parallel to the rays coming from Pollux.

Dick: How are Castor's rays focused?

Jane: All of the rays captured from Castor are focused at that same image plane, but at a different spot on the plane than the image of Pollux, because they come into the telescope from a direction that's three degrees away from Pollux. That's more of the "magic" of the shape of the lens. The location of Castor on the image plane is determined by the direction that its rays arrive at the telescope.

Dick: A lens is miraculous. Rays from a remote object are spread across the whole aperture, and jumbled together with rays coming from different remote objects, but the lens manages to sort them all out and map an image of all the remote sources onto a plane.

Jane: Yes.

Dick: Isn't a closer object, like a neighbor's house, focused at a different image plane, farther from the primary lens?

Jane: Yes, so at the same time the same amazing lens may also be receiving some diverging rays, which are also spread all the way across it, from, say a chimney protruding into the field of view. It simultaneously creates an image of the chimney, but at a different focal plane. But that's another subject. We were just talking mainly about why images of different objects are spaced apart on a particular image plane.

Dick: Ahh! Thanks for explaining all that, Jane. How do you happen to know about optics?

Jane: I work at WalMart as an optometrist.

Dick: Well I'm only an optimist. Would you like me to collimate your telescope? A Newtonian 'scope can have pretty poor images if it isn't collimated. It's worse than being unfocused.

Jane: Well I could ask Carl Kelley to do it. He collimates telescopes for people, using a laser.

Dick: May I do it for you?

Jane: Nooo. Well—uuummm. Oh, why not!

---

## Collimating Jane's Telescope

Dick: This is called the drawtube.

Jane: I know--that's the thing you look through. You don't have to tell me what I already know. I'm an optometrist.

Dick: Sorry.

Jane: So how do we go about collimating my telescope?

Dick: Jane, as you know, collimation is aligning the optical components to give the best image. There are several ways of doing it. I'll show you one.

Jane: How do we know it even needs it?

Dick: We'll test the telescope first by looking at a bright star. Let's look at Capella. I'll move the telescope to get Capella exactly in the center of the field of view.

Jane: OK.

Dick: Now we defocus Capella's image, so it gets much larger, until it becomes a large disc of dim light. Notice that there's a circular shadow in the disc—that's the shadow of the secondary mirror. Take a look.

Jane: I see shadows of the legs of the spider too.

Dick:: Yes, that's what those three radial dark lines are. If the telescope is well collimated, the mirror's shadow should be exactly in the center of the light disc.

Jane: It isn't in the center.

Dick: The defocused star isn't even a good circle: it's a weird crescent shape, so your telescope is pretty far out of alignment, Jane.

Jane:: Oh. Yes, I see that. Now what?

Dick: Well, let's collimate it. There are a lot of accessories we could buy to help, but we can make do without them. Let's start by making a dummy eyepiece. If you have an empty cassette from a roll of 35 mm film, we could punch a hole in the center of it and put it over the drawtube.

Jane: I don't have any empty film cassettes; I've gone digital.

Dick: OK, we can use a card with a small hole in it—a hole about six millimeters in diameter.

Jane: (Later). OK, here's the card. We do things like this in optometry.

Dick: We'll fasten the card over the eyepiece drawtube, with the hole exactly in the center. Now we need good lighting so we can see the inside of the telescope tube, so let's go inside.

Jane: (Later, peering into the little hole). What am I seeing on the outer edges?

Dick: If it's circular, that's the end of the drawtube. Now let's adjust the secondary mirror so it's outline is centered in that drawtube circle. While we're doing that, you don't have to worry about what's reflected in the secondary mirror.

Jane: (Later). OK, the secondary mirror is in the center.

Dick: Now let's look through the little hole again, at the reflection of the main mirror in the secondary mirror. If it isn't centered, we have to readjust the secondary some more until it is.

Jane: (Later). It's right in the center now.

Dick: OK, we're ready to adjust the main mirror. We have to adjust the three screws and the center screw until the outline of the end of the telescope tube is right in the center of the outline of the secondary, and so the mirror is snug and doesn't wiggle.

Jane: (Later). Phew! OK. That's done.

Dick: Now your telescope is collimated. You should get good images.

Jane: Where does the laser come in?

Dick: That's a different method of collimating.

Jane: Why didn't we use it?

Dick: I don't happen to own a laser collimator. Collimation is much easier with a laser, which costs only $50. You put a piece of paper over the aperture of the telescope tube, direct the laser beam into the eyepiece, and adjust the mirrors until you like what you see on the backlighted paper.

Jane: Thanks very much for doing all that, Dick.

Dick: You're welcome. Uuuh, Jane, there's going to be a solar eclipse that's visible from the Gulf of Mexico. Do you think you might want to see the eclipse from a cruise ship in the Gulf?

Jane: No! How long a cruise would it be?

Dick: Nine days, starting the 17th of next month.

Jane: Oh. (Pause.) I could get my own reservations. OK

# Resolving Power of Telescopes

Let's say we want to look at a binary star, which is a pair that revolves around a common center of mass under the influence of their mutual gravitational attraction. Gamma Centauri, Capella and Sirius are examples. About 57% of nearby stars have at least one companion star.

To observe, we go to a low-pollution site such as Arizona and set up our telescope. We need enough aperture to see both stars. The principle star is easy, being the brighter, but the companion may be a white dwarf that is only one one-thousandth as luminous, and very blue.

Studies of how binary stars orbit around each other are the only direct means of learning their masses. Double stars and their companions trace a corkscrew path against the sky. Sometimes the two are in line with the Earth, and it is difficult to separate their images.

We might see two objects as a single one if they are too close for the telescope to separate. Good resolution is crucial for studying binaries. Resolution or resolving power is the fineness of detail that can be distinguished in an image.

Here are some frequently asked questions.

How is resolution measured quantitatively? As an angle. It is the smallest angle that two objects can be apart without appearing to be a single elongated image.

What limits resolution? (a) atmosphere and (b) diffraction.

What is diffraction? The tendency of waves to bend around corners. Light, water and sound waves all diffract at the edge of an obstacle.

How does diffraction affect telescopes? When a parallel beam of light enters the aperture, diffraction spreads the light slightly, making it impossible to focus the beam to a sharp point even if the mirror is perfect.

What is an Airy disk? A point light source such as a star appears in a telescope as a disk with some dim circles around it. The image is named after George Airy, who calculated its diameter in 1834. Diffraction at a circular aperture makes a point source of light look like an Airy disk.

How does aperture size affect resolution? If two point sources are close enough together, their Airy disks may partially overlap, and the angle between them is the resolution limit. Resolution is inversely proportional to the diameter of the telescope. Practical limitations on the size of telescopes prevent improving resolution beyond a certain point.

Is resolution color-conscious? Blue light doesn't bend as much as red light at obstructions. The resolution limit due to diffraction is directly proportional to the wavelength. It is the same in music, where bass notes can be heard in neighboring rooms much better than high-pitched notes can.

How much effect does color have? With a 1-meter telescope under very good viewing conditions, blue light of wavelength 0.4 micrometers has excellent resolution, of only 0.1 arc second. With the same diameter of telescope, infrared light with a wavelength of 10 micrometers has 2.5 arc seconds resolution.

How do you compute the diffraction limit on resolution? About a hundred years ago British physicist Lord Rayleigh calculated that the resolution limit is 252,000 L/A arc seconds, where L is wavelength and A is the aperture (both in the same units such as meters). With a 4.5-inch telescope, a double star cannot be resolved if the separation is less than 1 arc second.

How troublesome is the limit? With the 36-inch Lick refractor telescope the minimum resolvable separation due to diffraction is 0.13 arc second.

Does the human eye have good resolution? Rayleigh's formula says the resolution of the eye, based on its dimensions, is about 20 arc seconds, but the least separation the unaided eye can actually resolve is much worse. A person with good eyesight can barely separate the stars of Epsilon Lyrae, which are 207 arc seconds apart. The discrepancy is due to the coarse structure of the retina.

---

## The Atmosphere's Mischief on Resolving Power

Michael goes out in the back yard and sets up his 5-inch Newtonian telescope. It isn't cloudy, the light pollution isn't too bad, and there's a new Moon. The telescope is well collimated and pretty solidly mounted on a tripod. Mirror and lenses are clean.

He starts looking at planet Mars, thirty degrees above the horizon. There it is, but it's fuzzy and slightly larger than it should be. There is a scale of "seeng" in which V is appallingly bad and I is perfect. It's called the Antoniadi scale of seeing, and this night Michael' seeing is only IV.

Here's the male fantasy part: A beautiful fairy suddenly appears out of nowhere and offers to show him what's causing the poor resolution. She spreads her wings and lifts Michael heavenward into the stratosphere. Then they follow a flight path that accompanies the image of Mars as it comes down through the Earth's atmosphere to Michael's telescope.

Fairy Lady explains that small regions of differing refractive index in the atmosphere move about and cause the direction of the light to change very slightly, making the atmosphere like a fluctuating lens. Scintillation occurs as the light passes through the air, like the shimmering above a hot road. As a result, in Michael's telescope an image wanders rapidly around its average position. In the case of a wide light source like a planet, scintillation makes the image's features and outline fuzzy.

The 5-meter Hale telescope doesn't resolve any better than a 10-cm telescope because of atmospheric distortion (but of course the Hale is able to detect much dimmer objects). Under the very best seeing conditions the atmosphere imposes a resolution limit of only 0.35 arc second, but that's not in a back yard in Cleveland, Ohio.

Of course space telescopes don't have atmospheric problems at all. An Earth-based telescope's image of the large Andromeda Galaxy, if done with a resolution of 10 arc seconds, is very fuzzy, but Hubble images show the features with spectacular clarity.

Fairy Lady then tells Michael about adaptive optics, which he had already read about. Atmospheric disturbances are monitored by observing a bright reference star (or a beam) and correcting the primary telescope to compensate for the scintillation of the reference star. A small thin mirror in the telescope is continually deformed, within hundredths of a second, to cancel the jiggling.

Michael would love to obtain adaptive optics, but that requires buying some new equipment that he can't justify as an amateur.

More fantasy part: The beautiful fairy then gives Michael the necessary adaptive optics equipment as a reward for finding a lot of stars of the Messier catalog last year. She pats him on the head and POOF–disappears in smoke. He sets up the new gear and focuses again on Mars. Atmospheric distortion is almost eliminated; the resolution is limited only by Rayleigh diffraction.

---

## Adaptive Optics

Dr. Kosmo, astronomer: The European Extremely Large Telescope (E-ELT) is corrected with laser adaptive optics to reduce the distortions of atmospheric turbulence. Under the very best seeing conditions the uncorrected atmosphere imposes a resolution limit of 0.35 arc second, but it usually isn't nearly that good.

Of course space telescopes don't have atmospheric problems at all. The problem with space telescopes is they are very expensive.

Stella: How good is the adaptive optics technique?

Dr. Kosmo: With adaptive optics turned off, an object may appear to be a single star with a badly blurred image. With adaptive optics turned on, it might be very clearly seen to be a binary star–two stars dancing in a circle.

Stella: That's so romantic, like dancing with the stars!

Dr. Kosmo: Uuh, yes. And they are both stars.

Stella: How does the adaptive optics system on the European Extremely Large Telescope work?

Dr. Kosmo: The E-ELT looks at one (or more) bright reference star, either a real star or an artificial one, that is in the same direction as the real object of interest. The actual location of the reference star or beacon is known pretty accurately. When the image of the reference star jiggles, a computer recognizes that the jiggling is caused by turbulence in the atmosphere.

The target that is the real object of interest is probably jiggling in the same way as the reference star, because it passes through the same part of the atmosphere. The computer quickly corrects the image of the real object of interest by the amount and direction that the reference star appears to jiggle.

Corrections made by adaptive optics increase both the spatial resolution and the sensitivity of the telescope. Unfortunately, Rayleigh diffraction still limits the resolution; it is inversely proportional to aperture size, and occurs over the full aperture of large optical and infrared telescopes.

Stella: The idea is clever. How does the telescope make the correction?

Dr. Kosmo: The telescope has deformable mirrors that are in the light path of the telescope. The computer creates compensating adjustments in the shape of. those mirrors, which are small and thin. They have thousands of actuators, and work in hundredths of a second, which is faster than the atmosphere changes.

Stella: What kind of reference star does the computer use?

Dr. Kosmo: The E-ELT uses natural stars and also creates its own artificial star(s) at about 60-mile altitude. The artificial star(s) are created by using several sodium lasers located around the telescope. The lasers excite sodium atoms in the mesosphere and thermosphere and cause them to glow.

Adaptive optics is already important in astronomy and will be used a lot more in the future.

## Spectrographs and a Family of Astronomers

Dad: Hi, Brad.

Brad: Hello, Mr. Hanson.

Mom: Celeste tells us that you're a classmate of hers in 11$^{th}$-grade science.

Brad: Yeah.

Celeste: Brad has to do a paper on spectrographs. I told him you and mom do something with them at work.

Dad: Yes–we know a little bit about them.

Brad: What is a spectrograph anyway?

Mom: Brad, you probably already know that light can be made up of many colors, and that they can be spread out into a spectrum.

Brad: Yep. We had that. Newton used a prism to spread it out, because different wavelengths are refracted in slightly different directions.

Dad: Well, a spectroscope has a slit that lets in light from a source, and the light coming out of the slit passes through a lens and a prism, so it produces a pattern that looks like a bar code.

Mom: Each bar on the pattern is nothing but a focused image of the slit in a different color. If the pattern is recorded, the instrument is called a spectrograph. If the intensities of the lines are measured, it is called a spectrophotometer.

Brad: How do they record the light or measure its intensity?

Dad: At first it was done chemically on a photographic emulsion. A second slit can isolate any particular line. In recent years, the lines are instead measured electronically with charge-coupled devices, abbreviated CCD's. They are connected to a computer.

Mom: Spectrographs can be made for visible light, near-infrared, and near-ultraviolet and other wavelengths.

Celeste: The teacher also said something about a diffraction grating spectroscope.

Dad: Yes. That's a different type. It uses a diffraction grating instead of a prism to fan out the light into its individual colors. The diffraction grating is a metal or glass mirror or a lens on which a lot of parallel lines have been inscribed by a diamond. Sometimes the lines are cut on a concave mirror instead of a flat piece, so that the grating also focuses the light, and lenses aren't needed.

Brad: What are these things good for?

Mom: Chemists analyze spectra to identify chemical elements. Different atoms, molecules and ions produce characteristic lines and bands in emission and absorption spectra. Chemists compare the spectrum of an unknown specimen with known spectra of elements such as sodium.

Dad: Also, astronomers use spectroscopes to study the emission and absorption spectra from stars and other celestial objects. If a spectroscope is mounted on a telescope it is usually at the Cassegrain focus. The Doppler effect or the stretching of space can shift the wavelengths of an element's characteristic spectrum, for example, toward the red. The amount that the wavelengths have been shifted lets astronomers estimate the relative speed, at least the radial component of speed, of a star that's emitting the radiation.

Mom: When the comet Shoemaker-Levy crashed on Saturn in the 1990s, some heated gases came to the surface, and astronomers analyzed them with spectrographs and spectrophotometers to learn more about the composition of Saturn and its atmosphere.

Celeste: OK. My brain is getting fried. Let's go to Malley's for ice cream, Brad.

Brad: Yeah. I know enough now to get by. Thanks, Mr. and Mrs. Hanson.

## Signals and Ubiquitous Noise–S/N Ratios

Rick didn't hear Marie thank him for helping with the housework because vacuum cleaner noise corrupted the message. Noise is everywhere, and there are all kinds, not just acoustical. Noise is false or irrelevant background, stacked on the useful information. Financial risk managers speak about noise in stock-picking systems. It is involved in medicine, biology, the weather service, and electronic warfare. Some religious writers even refer to a differing interpretation of the Bible as "noise" that distorts the basic message.

Signal-to-noise ratio (S/N) is the ratio of the average power in a signal to the average power in the noise. The ratio of signal to noise is a factor in designing digital cameras; a S/N ratio of at least 5 is needed to distinguish image features with high fidelity. There is a threshold of S/N ratio below which any particular channel is useless.

Many optical telescopes have CCDs (charge-coupled devices). A CCD is a light-sensitive electronic detector invented in 1970 that is widely used in ground- and space-based astronomy for imaging, photometry and spectroscopy. CCDs are usually flat and as small as several square centimeters, so they don't have a large field of view like photographic plates.

Incoming photons that strike the CCD release electrons that are collected from the pixel addresses on the CCD and counted. However, noise is coming in along with the desired signal. More noise is introduced when charges are moved out of the CCD, amplified, digitized, and stored in a computer, which puts a lower limit on the signal that can be accurately recorded.

Noise is a bugaboo of the sub-field called radio astronomy. Radio telescopes are large antennas for recording and measuring the radio-frequency emissions from celestial radio sources. Starting 75 years ago, radio telescopes have identified many sources of radiation, including cosmic microwave background, radio galaxies, pulsars and masers.

It is better for a speaker to let the audience miss something than to bore them, so it is often all right to skip redundant statements even at the cost of misunderstanding. Although redundancy in speaking may seem to be disrespectful "talking down," in astronomy redundancy is good; it is invaluable in extracting signals from noisy backgrounds.

Assume that the signal-to-noise ratio of a data transmission is 2 and the maximum permissible error in the message is 3%. For a S/N ratio of 2, a certain amount of redundancy is required to limit the error to only 3%. Redundancy improves accuracy. The required redundancy can be calculated using Norbert Wiener's information theory. Wiener was a famous American child prodigy–a mathematician and philosopher who died in 1964 at age 69. A PhD from Harvard at age 18, he pioneered noise theory and many other topics.

Since most noises are constantly varying and many astronomy signals are systematic, an astronomer can sometimes improve a S/N ratio by observing for a longer time and correlating the currently received signals with prior signals, or by observing an object simultaneously from two stations of differing noise, then superimposing and correlating the data.

---

## What's the Difference–?

1. Between a Schmidt Telescope and a Super Schmidt Telescope? Schmidt telescope (1930) has a spherical mirror, with a transparent correcting plate in the incoming light beam before it reaches the mirror, to correct for spherical aberration. An uncorrected spherical mirror has severe spherical aberration, but no coma. A paraboloid has severe coma but no spherical aberration. Spherical aberration causes light arriving near the periphery of a mirror to focus at a different distance than light arriving near the center of the mirror, so the light does not all focus at a single point. Coma, on the other hand, causes each toroidal zone of the mirror to produce a spurious image in the form of a circular patch of light around an off-axis star.

Because of its correcting plate, a Schmidt has neither much coma nor spherical aberration. It focuses the light well over a very wide field, provided it is focused onto a curved surface. If there is a recording detector such as a photographic plate, it must be sprung over the curved surface.

Super Schmidt A Super Schmidt telescope (developed by American James G. Baker about 1958) has additional correcting plates that retain the wide field while greatly increasing the speed so that cameras can record meteor and satellite trails. Its design is complicated.

2. Between Revolution and Rotation? No difference, really. But revolution seems to be better for describing the travel of an entire body in an orbit around another body, and rotation is often used to describe the spinning of a body on an axis that passes through the body.
3. Between an Amateur Astronomer and a Professional? A professional gets paid for it but amateurs do it because they love to.
4. Between Cassini and Casini? In 1659 Huygens discovered a ring around Saturn. In 1675 astronomer Giovanni Cassini realized that what Huygens thought was one ring was actually two, with a gap between them. "Cassini's division" separates the B and A rings of Saturn. Nowadays, the Huygens/Cassini project is a space expedition to explore Saturn.

The astronomer Giovanni Cassini is not your dad's Giovanni Casini, who was a great music composer, organist and scholar. They were contemporaries in Italy.

5. Between a Telescope, a Microscope, and a Periscope.
   Tele. = Up, far, big.
   Micr. = Down, close, small.
   Peri. = Bent, near, bad.

References:

For "Signals and Noise"

1. Dictionary of Astronomy. Facts on File, Infobase Publishing. New York NY. 2006.

2. "Signal-to-Noise Ratio." Dictionary, Onelook.com.
3. "Signal-to-Noise Ratio." Radio Astronomy. Norbert Wiener. wikipedia.org
4. "Signal-to-Noise Ratio." Astronomy.net. God & Science Forum Message. Mar., 2000.

# Chapter 16
# Pronunciation and Definitions

## Pronunciation of Astronomy Words

You are probably already pronouncing most of these words correctly, but perhaps not all.

Betelgeuse A remote luminous red supergiant that is the second-brightest star in the constellation Orion. Rhymes with beetle jooz," with accent on the first syllable, and a soft g.

Bootes A large constellation in the northern hemisphere near Ursa Major, and whose brightest star is Arcturus. Three syllables. Bootes rhymes with "go oat ease," with accent on the second syllable.

Christian Huygens Dutch mathematician, physicist and astronomer, 1629-95. Rhymes with "high gens," with accent on the first syllable, and a hard g in the second syllable.

Charles Messier French astronomer who prepared a star catalog printed in final form in 1784. Messier has three syllables, and rhymes with "messy ay." The final r isn't pronounced.

Pleiades A loose cluster of stars in the constellation Taurus, six of whose stars are visible to the naked eye. Has three syllables, and almost rhymes with 'three a these."

Tycho Brahe A great astronomer who in 1580 proposed an erroneous model of the solar system. Rhymes with "peeko bra hee."

Uranus The planet that is seventh in order from the Sun. Almost rhymes with you're a mess." Main accent is on the first syllable.

---

## More Astronomy Words to Pronounce

Student: "The astronomy subject I'm going to give a talk about this semester is supernovas."

Professor: "I'm sorry, Miss Blaze, but you've lost two points already because the preferred plural of supernova is supernovae (pronounced 'supernovee')."

Student: "You're taking off points just because of my pronounciation?"

Professor: "Yes, and incidentally, the word is pron<u>u</u>nciation, not pron<u>ou</u>nciation.

You won't lose any points for that because this is not an English class."

Student: "With all due respect, Dr. Nadir, nobody loves a grammarian."

Professor: "I don't like it either, but it's just to help you get a job."

Student: "Are there other astronomical terms that are often mispronounced?

Professor: "Yes indeed. Here are some of them, with phonetic spelling and the accented syllables capitalized.

Uranus YOUR-an-us Seventh planet from the Sun.

Aphelion a-FEEL-yon Point farthest from the Sun on an orbit.

Tycho Brahe TEE-ko-BRAW-hee Meticulous Danish astronomer.

Cepheid   SEE-fee-id   Regularly pulsating star.

Nebulae   NEB-you'll-ee   Plural of nebula, a galaxy other than Milky Way.

Nuclei   NOO-clee-eye   Plural of nucleus. Small, brighter, denser portion.

Betelgeuse   BEET-el-joos   Red giant star near Orion's shoulder.

Ephemeredes   eff-e-MARE-e-dees   Plural of ePHEMeris, a table of locations of a body.

Autumnal   Aw-TUM-nal   Relating to the autumn season.

Globular cluster   GLOB-you-lar   A star cluster shaped like a globe.

Io   EYE-oh   One of Jupiter's Moons.

Isotropy   eye-SAW-tro-pee   Noun. Property of being the same in every direction. But the adjective is pronounced eye-so-TROP-ic.

Troposphere   TROW-po-sphere   Lowest layer of Earth's atmosphere.

Charon   CAR-on   Satellite of Pluto. Rhymes with carbon.

Homogeneous   ho-mo-GEEN-ee-us   Of uniform kind throughout.

Joule   JOO-el   A unit of energy.

Longitude   LON-je-tood   Angular distance from a reference meridian.

Kilometer   kill-OMM-meter   Metric unit of distance. (Doesn't rhyme with centimeter.)

Zodiacal light   zo-DIE-a-kel   A diffuse glow seen in the west before twilight and in the east before dawn.

Student: "Thanks, Dr. Nadir. Should I give my report about supernovae now?

Professor: "Yes please, Miss Blaze. And I don't want anyone to hate me, so I won't take off the two points."

# Definitions

Halley's Comet

a) A comet recorded as early as 240 BC, which has an average period of 76 years, and whose orbit was calculated by Edmond Halley in about 1740 by using Newton's laws.

b) A popular rock band.

Michelson and Morley

a) Case and Western Reserve scientists whose experiments showed that electromagnetic waves do not require any medium such as ether in which to propagate.

b) Men who first compared the speed of light straight ahead in the direction of the observer's travel with the speed at a right angle to the observer's travel.

Lyre or Lyra

a) A constellation in the northern hemisphere lying partly in the Milky Way, and including the star Vega.

b) A stringed instrument of the harp family used to accompany a singer or reader of poetry, especially in ancient Greece.

Perturbation

a) A small change from regular behavior, for example the deviation of a planet from an elliptical orbit due to the gravitational effects of other planets.

b) Reaction to something you didn't like.

Microwave Background Radiation

a) Radiation that still remains from the primordial fireball of the Big Bang. It was predicted theoretically in 1940 and 1948, and discovered experimentally in 1965.

b) Leakage from an electronic oven.

Local Group

a) A small cluster of thirty to forty galaxies of which our Galaxy and the Andromeda galaxy (M31) are prominent members.

b) Cuyahoga Astronomical Association.

Great Wall

a) The largest coherent large-scale structure detected so far. It was detected in redshift surveys of galaxies, and is about 310,000 light years away. A structure consists of very long filaments of galaxies surrounding huge voids that are empty of matter.

b) A line of fortifications extending about 1500 miles across northern China, built in the third century by criminals, conscripted soldiers and slaves.

Dish

a) An antenna used in radio telescopes and consisting of a large spherical or parabolic metal reflector for bringing radio waves to a focus at a point where a "feed antenna" is located to pick up the signal.

b) An attractive person, e.g., Britney Spears.

Calypso
a) A small irregularly shaped satellite of Saturn, discovered in 1980.

b) An orchid having a rose-pink flower with an inflated pouch lip usually marked with color.

Albedo
a) The reflecting power of a non-luminous object such as a planet.

b) Owner of Bedo Cement Contractors.

Cannibalism
a) The swallowing of one galaxy by a larger galaxy, especially in the center of a cluster of galaxies; cannibalism probably contributed to growth of the most massive galaxies.

b) The practice of humans eating human flesh, or of animals eating creatures of their own kind.

All of the above answers are correct, so everyone gets them all right.

# About the Author

Charles Grace is an engineering manager and lawyer. His early astronomy training was with Dr. Jason J. Nassau at Case Institute of Technology. Charles Grace is an honorary life member of the Cuyahoga Astronomical Association, a member of the Cleveland Astronomical Society and a member of many professional organizations and honorary fraternities.

He has a Doctor's Degree in Electrical Engineering from Carnegie Institute of Technology and a Juris Doctor degree from Cleveland State University. Dr. Grace worked fifteen years as an engineering manager and consulting engineer and fifteen years as a supervisory attorney. His interests besides astronomy are music, parliamentary law, books and public speaking. He lives in Westlake, Ohio.

# Index

**D**



**E**

**I**



**J**



**K**



**L**



**M**

**S**



**T**

**U**



**V**



**W**



**X**



**Z**